EQUIPMENT MANAGEMENT

Key to Equipment Reliability and Productivity in Mining

Second Edition

Paul D. Tomlingson

Published by

Your most precious resource.

Society for Mining, Metallurgy, and Exploration, Inc. (SME)
8307 Shaffer Parkway
Littleton, Colorado, USA 80127
(303) 948-4200 / (800) 763-3132
www.smenet.org

SME advances the worldwide mining and minerals community through information exchange and professional development. SME is the world's largest association of mining and minerals professionals.

Copyright © 2010 Society for Mining, Metallurgy, and Exploration, Inc.

All Rights Reserved. Printed in the United States of America.

Information contained in this work has been obtained by SME, Inc., from sources believed to be reliable. However, neither SME nor the authors guarantee the accuracy or completeness of any information published herein, and neither SME nor the authors shall be responsible for any errors, omissions, or damages arising out of use of this information. This work is published with the understanding that SME and the authors are supplying information but are not attempting to render engineering or other professional services. If such services are required, the assistance of an appropriate professional should be sought.

No part of this publication may be reproduced, stored in a retrieval system, or transmitted in any form or by any means, electronic, mechanical, photocopying, recording, or otherwise, without the prior written permission of the publisher. Any statement or views presented here are those of the authors and are not necessarily those of SME. The mention of trade names for commercial products does not imply the approval or endorsement of SME.

ISBN-13: 978-0-87335-315-1

Library of Congress Cataloging-in-Publication Data
Tomlingson, Paul D.
Equipment management : key to equipment reliability and productivity in mining / written by Paul D. Tomlingson. -- 2nd ed.
p. cm.
Rev. ed. of: Equipment management: breakthrough maintenance management strategy for the 21st century / Paul D. Tomlingson. 1998.
Includes bibliographical references and index.
ISBN 978-0-87335-315-1
1. Mine management--Handbooks, manuals, etc. I. Tomlingson, Paul D. Equipment management. II. Title.
TN274.T55 2010
622.068'2--dc22

2009040762

Contents

Introduction

The mining and mineral processing industry is among the most difficult working environments in the world. Sheer rock is blasted, moved, crushed, and processed to yield the everyday indispensable commodities often taken for granted. The equipment that make these processes possible are massive, powerful, complex, and difficult to maintain. Maximum equipment reliability is required as the worldwide demand for mined products increases along with the requirement to produce them more efficiently.

Equipment reliability is the theme of this book. But the maintenance organization is no longer the sole focus of this task. That role has been shifted to the mine manager and his or her ability to bring all resources successfully into the task. The maintenance discipline has never been a stand-alone activity. Its success depends on the mine manager creating an environment in which maintenance can succeed. This is accomplished by converting the broad corporate mission into a production strategy that focuses all of the manager's resources on their interdependent efforts to ensure that the operation has the consistently reliable production equipment necessary to make the operation profitable.

Development of the mine manager's production strategy is guided by the principles of equipment management. Specific, interlocking and mutually supporting objectives are assigned to each department with maintenance-related responsibilities inherent in each. Then policies are provided to amplify the exact way in which individual departments are expected to operate internally and interact with all other departments to ensure the desired equipment reliability. In response to these policies, individual departments develop or revise internal and external interactive procedures. Collectively, these procedures comprise the programs that each department is expected to carry out. But the programs specify both internal and interdepartmental procedures to ensure a mutually supporting outcome. The maintenance program, for example, would prescribe internal procedures such as scheduling preventive maintenance services as well as interdepartmental procedures such as obtaining warehouse materials. Thus, equipment management ensures that maintenance has the support and cooperation needed to be successful and that the desired production equipment reliability is realized.

Equipment management does not stop there. Based on the specific equipment needs of the operation, modern reliability strategies are added to the departmental programs so that pertinent technologies can be incorporated into the effort. Then the best information is applied to control the overall effort against specific performance goals. Finally, regular, all-encompassing evaluations are prescribed to identify improvement opportunities and convert them into realities. The effective mining manager oversees these collective efforts which have been made logical and manageable by following the principles of equipment management.

CHAPTER 1

Understanding Equipment Management

WHY IS IT NECESSARY?

The main objective of equipment management is to create an effective maintenance capability to meet the increasing reliability demands of complex modern production equipment. For too long, maintenance has been seen as a last resort for restoring abused equipment rather than a key contributor to effective plant operation. Maintenance departments cannot change these circumstances by themselves. Change must be initiated by enlightened mine and plant managers who understand the potential of quality maintenance. But maintenance must change as well. It must master new technologies, organizations, and management skills. Beneficial change must be a total plant objective. Only then can the intensified reliability demands of modern industry be successfully met. The equipment management strategy guides this essential change.

Modern industry has entered an era when maintenance can no longer be a "necessary evil" characterized by stubborn adherence to past practices such as rigid craft organizations or promotion based only on longevity. Along with greater global completion, increasingly complex modern production equipment has become complex and more difficult to maintain. It requires today's maintenance managers to balance the technical aspects of effective maintenance with the leadership demands of a proactive work force. Supervisors are no longer the "folk" heroes who could get any broken equipment running again. They are now managers of professional craftsmen who have mastered reliability technologies to preclude the far-reaching consequences of equipment failure. Overseeing this transformation is an enlightened plant manager who has created an environment in which maintenance will be successful and the plant profitable. These plant managers have provided guidance to ensure that maintenance is an integral part of the plant's operating plan. These are the managers whose plants will be profitable. Their maintenance departments will have the full support and cooperation of all other departments (Table 1-1).

TABLE 1-1 Control of critical maintenance elements

Priority	Control Element	Influence Rating	Degree of Maintenance Control, %	Control Index	Primary Source for Improvement
1	*Labor productivity*	*10*	*60*	*6.0*	*Maintenance*
2	Material control	10	20	2.0	Other
3	*Leadership*	*9*	*70*	*6.3*	*Maintenance*
4	Work load	9	30	2.7	Other
5	Organization	8	50	4.0	Other
6	Interdepartmental relations	8	20	1.6	Other
7	Cost data	7	20	1.4	Other
8	Performance data	7	50	3.5	Other
9	*Preventive maintenance procedures*	*7*	*75*	*5.3*	*Maintenance*
10	Planning	6	60	3.6	Other
11	Scheduling	5	50	2.5	Other
12	Training	4	80	3.2	Other
13	Maintenance engineering	4	40	1.6	Other
14	Technology	3	90	2.7	Other
15	Labor practices	2	20	0.4	Other
Maintenance control of critical elements			49		

NOTE: To help establish the premise that successful mining maintenance requires cooperation and support across the total operation, 21 different domestic and international mining operations were surveyed. Maintenance managers rated the relative importance of 15 vital maintenance control elements and listed them in an order of priority. Next they rated the degree of direct influence they had over each element (10 being highest). Then they rated the degree of control (%) that maintenance had over each element. These two ratings were then multiplied to yield a control index. The results indicated that maintenance could substantially influence only 3 of 15 control elements (*italicized index value of more than 5.0).* The remaining 12 elements could only be improved with support from other departments. Overall, maintenance had only 49% control over their destiny. Successful maintenance would require support from the total operation.

PLANT MANAGEMENT ROLE

Prominently displayed company mission statements address broad goals for producing quality products, achieving customer satisfaction, utilizing skilled employees, providing good working conditions, and operating profitably. But mission statements say nothing about maintenance; thus the need for the plant manager to translate the company mission into a production strategy for his plant. That strategy will acknowledge maintenance as an activity requiring strong management support, solid cooperation from operations, and first-rate service from staff organizations. Equipment management emphasizes these unifying attributes by guiding development of a production strategy using meaningful department objectives amplified with policies so that individual departments can develop day-to-day procedures to conform to the policies. In turn, these collective procedures become departmental programs prescribing internal and interdepartmental actions. The result of the enlightened production strategy is a well-understood maintenance program supported by quality information. Then, in recognition of the intensive reliability demands of complex, modern production equipment, selected reliability centered maintenance and total productive maintenance attributes are added to the basic maintenance program. Collectively, these actions better ensure reliable equipment yielding full productive capacity and better assurance of a profitable operation.

MAINTENANCE PROGRAM

An effective maintenance program is the direct by-product of a well-conceived production strategy. It describes the interaction of all departments and individuals as they request or identify work; classify it to determine the best reaction; then plan (as required), schedule, assign, control, and measure the resulting work; and finally, assess overall accomplishment against goals such as performance standards and budgets. It documents internal procedures such as how to plan a major job, but it also explains interdepartmental actions such as scheduling work with operations or obtaining materials from the warehouse. The program exists as much to ensure well-ordered procedures within maintenance as to advise other departments how to help maintenance carry out their services effectively.

Unfortunately, adequate program definition has been the most neglected yet essential task of maintenance. As a result, there is unnecessary confusion within maintenance, poor utilization of maintenance resources by confused operations departments, and unsatisfactory support by material control departments who try to second-guess what maintenance wants but never specifies. Plant managers are not only disappointed with the results but frustrated about how to correct the situation. To illustrate,

- Of 31 maintenance organizations evaluated between 1986 and 2001, none had a well-documented maintenance program. Only six maintenance managers could explain what the total program was "supposed to do." Few of their subordinates were able to fully describe every element of the "program." Many thought that the newly purchased information system was the program.
- Of the 31, 20% attempted self-directed teams staffed with craftsmen. All failed because a properly documented program providing work control procedures had not been provided. Although these craftsmen were well qualified to diagnose and repair equipment problems, none had been educated on the work control procedures previously provided by their supervisors. The existence of a well-defined program and education about its procedures could have avoided these failures.

- Of 13 heavy industry maintenance clients evaluated during the same period, nine did not have a program in place and needed consulting support because they had purchased an information system that they were incapable of using. The information system only added confusion and frustration. The remaining four clients were encouraged to pause, document their program, educate personnel, and then resume use of the new information system. These operations made a significant recovery and profited by the use of quality information.

The absence of a well-defined and well-understood maintenance program is the direct cause of numerous serious maintenance management problems and unsatisfactory results. But the absence of the objectives and policies on which the program could be built is often the primary reason for not having a program as well as the problems that follow. Any plant manager disappointed with maintenance performance must first ask himself whether the lack of a production strategy with its objectives and policies is an underlying cause.

The maintenance program is discussed in subsequent chapters.

MUTUALLY SUPPORTING DEPARTMENT OBJECTIVES

Objectives assigned to departments by the production strategy ensure the harmonious interaction of all plant departments. They reinforce the idea that successful maintenance is built on interdepartmental cooperation and support. Equipment management might broaden the basic maintenance objective (defined in the next section) to prescribe mutually supporting, interactive working relationships, acknowledging the need for cooperation between departments to ensure maintenance effectiveness (Table 1-2).

The expected result of these objectives is the harmonious interaction among departments. When maintenance presents a work schedule, operations accepts the obligation to comply with the approved schedule. Plant managers can never assume that "someone who has been in this plant for 20 years should already know this." The production strategy precludes this often serious mistake.

THE MAINTENANCE OBJECTIVE

The objective assigned to maintenance states the principle tasks. It links maintenance with other key departments, ensuring their mutual support of the plant production strategy. Typically,

> The primary objective of maintenance is *to keep production equipment in a safe, effective, as-designed operating condition so that production and quality targets can be met on time and at the least cost.*
>
> A secondary objective of maintenance is to perform approved, properly engineered, and correctly funded nonmaintenance work (such

TABLE 1-2 Typical working relationships

Interaction	Actions
Maintenance to operations:	Advise of program details and procedures Provide preventive maintenance (PM) services Replace major components Make emergency repairs Plan and execute selected major jobs Perform overhauls Advise of required maintenance Train operators on PM-related tasks Provide information on repair decisions
Operations to maintenance:	Understand and support the program Advise of production schedules Specify availability of equipment Require operators to perform PM checks Operate equipment correctly Report problems promptly Comply with approved maintenance schedules Require quality service Stay informed of cost and performance

as construction and equipment installation) to the extent that such work does not reduce the capability for carrying out the maintenance program. In addition, maintenance will operate support facilities (such as power generation) but will ensure that necessary resources are allocated within its authorized work force and are properly budgeted. As appropriate, maintenance will also monitor the satisfactory performance of contractor support when utilized to perform maintenance or capital work.

A primary as well as a secondary objective are provided to establish a clear precedence. That is, production equipment maintenance is first while project work follows, resources permitting. Further, the use of phrases like "as designed" means that equipment modification is excluded from the primary maintenance task (it is not maintenance). This implies that all nonmaintenance work is to be approved, properly engineered, and correctly funded. Such work may only be done if it does not reduce the capability of maintenance to meet its primary function. If maintenance is to perform operating functions such as power generation, it must be properly staffed and budgeted to do so. With an objective such as that cited, there should be little question about maintenance priorities and limits as to what and how much it can do. Based on a clear objective, maintenance personnel know their responsibilities exactly and can organize properly to carry them out. Their customers are also aware of maintenance limitations and will request support accordingly. The objective also helps to ensure that the intended maintenance program can be carried out effectively. The absence of a clear objective can yield unfortunate consequences.

TABLE 1-3 Typical department objectives and goals

Department	Objective	Goal
Purchasing	Provide purchasing and contracting support to obtain materials and services as requested by operating departments. Permit maintenance, maintenance–support, or projects to be dependably scheduled and carried out with on-time completion at agreed-upon costs.	85%
Warehousing	Stock and replenish specified repair materials, components, and consumables to ensure they are available for plant use as required. Arrange procedures to have selected components rebuilt and restocked in inventory. Provide effective issue and return procedures.	85%
	Operate tool room to ensure availability and accountability for specified tools.	100%
Accounting	Establish a suitable information system that allows operating departments to develop and utilize information to control operations and work while providing plant-level cost and performance information. System should also provide for control of inventory and purchasing activities. Confer with all departments as system is developed or identified. Ensure system emphasizes ease of use and highest capability among field personnel to develop data as a basis for timely, accurate, and complete information.	85%
Operations	Operate equipment properly to meet established production, quality, and cost targets. Utilize maintenance services to help ensure reliable equipment. Incorporate operator maintenance in conjunction with maintenance. Observe guidelines in requesting nonmaintenance support. Follow established work order procedures in requesting work and utilize it to control work performed by operating personnel. Conduct weekly scheduling meetings with maintenance and engineering to determine the requirement for equipment shutdown for the coming week. Negotiate best shutdown times to comply with needs.	85%

Objectives such as those shown in Table 1-3 are typical tasks assigned to departments in the equipment management strategy, and they are mutually supporting. A task assigned to one department (conducting preventive maintenance [PM] services) has a corresponding action by cooperating with the other department (operations makes equipment available).

Clear departmental objectives minimize the surprises that some seasoned plant managers have experienced by assuming maintenance needs no guidance. The following situation is typical:

> In a large processing plant, the 300-employee maintenance work force was having difficulty. The backlog of incomplete work was rising sharply, emergency repairs were excessive, and downtime was unacceptable. During shutdowns, they needed contractors' help to catch up. Investigations of daily work showed that 30% of their work was not maintenance but equipment changes, modifications, or installations. Much of it was being done without plant engineering

knowledge. Nearly 35% of their labor was used on this work, detracting from labor required for basic maintenance. When the plant manager realized this situation, he clarified the maintenance objective and educated personnel in the proper use of the maintenance work force. Within a year, no contractors were needed, the backlog was under control, and emergency repairs were infrequent.

POLICIES AMPLIFY OBJECTIVES

Policies contained in the production strategy are primarily intended to amplify department objectives to preclude any misunderstanding of departmental roles and responsibilities. In turn, policies are the basis for developing day-to-day procedures for interactions between departments. For example,

> The overall PM program will be assessed annually to ensure it covers all equipment requiring services and that the most appropriate types of services are applied at correct intervals. The performance of the PM program in reducing equipment failures and extending equipment life will be verified.

Such policies not only make maintenance responsible for establishing the PM program but specify how it is to be conducted in the best interests of the plant. Policies like these are clear evidence that the plant manager understands the importance of good preventive maintenance and has taken positive steps to ensure its effectiveness.

SUGGESTED MAINTENANCE POLICIES

Typical maintenance policy guidelines are suggested in the following sections.

Department Responsibilities

- Department managers will ensure compliance with the policies for the conduct of maintenance.
- Each department will develop and publish procedures by which other departments may obtain their services.
- Operations will be responsible for the effective utilization of maintenance services.
- Maintenance will be responsible for developing a pertinent maintenance program, educating personnel on its elements, and carrying it out diligently. It will also make effective use of resources and ensure that quality work is performed.
- Maintenance will publish work load definitions and appropriate terminology to ensure their understanding and proper utilization.

Preventive Maintenance

- Maintenance will conduct a detection-oriented PM program. The program will include inspection, condition monitoring, and testing to help uncover equipment deficiencies and avoid premature equipment failure. The PM program will also provide lubrication services, cleaning, adjusting, calibration, and minor component replacement to help extend equipment life.
- Preventive maintenance will take precedence over every aspect of maintenance except bona fide emergency work.
- No major repairs will be initiated until PM services have established the exact condition of the equipment and elements of the repair have been correctly prioritized.
- The overall PM program will be assessed annually to ensure it covers all equipment requiring services and that the most appropriate types of services are applied at correct intervals. The performance of the PM program in reducing equipment failures and extending equipment life will be verified.
- Equipment operators will perform appropriate PM services to help ensure the reliable operation of equipment.
- Compliance with the PM schedule will be reported to management and supervisors controlling "no show" equipment identified.

Planning and Scheduling

- Criteria will be established to determine which work will be planned.
- Planning and scheduling will be applied to comprehensive jobs (e.g., overhauls, major component replacements) to ensure that work is well-organized in advance, properly scheduled, and completed productively and expeditiously.

Priority Setting

- Maintenance will publish a priority-setting procedure that allows other departments to communicate the importance of work and maintenance to effectively allocate its resources.
- The priority-setting procedure will specify the relative importance of jobs and the time within which the jobs should be completed.

Information

- Maintenance will develop and use information concerning the utilization of labor and the status of work, backlog, and cost and repair history to ensure effective control of its activities and related economic decisions such as equipment replacement.

- The work order system will be used to request, plan, schedule, assign, and control work.
- Minimum necessary administrative information will be developed and used.
- Performance indices will be used to evaluate short-term accomplishments and long-term trends.

Organization

- The maintenance work load will be measured on a regular basis to help determine the proper size and craft composition of the work force.
- The productivity of maintenance will be measured on a regular basis to monitor progress in improving the control of labor.
- Every effort will be made to implement and utilize organizational and management techniques to ensure the most productive use of maintenance personnel.

Maintenance Engineering

- Maintenance engineering will be emphasized to ensure the maintainability and reliability of equipment.
- Reliability technology will be utilized to facilitate effective maintenance.
- No equipment will be modified without the concurrence of maintenance engineering.
- All new equipment installations will be reviewed by maintenance engineering to verify work quality and maintainability.

Material Control

- Established procedures for purchasing and the withdrawal or return of stocked materials will be strictly adhered to.
- No components or parts will be removed from any unit of equipment and used to restore another unit to operating condition unless authorized by the maintenance manager.
- Maintenance will promptly return unused stock materials to the warehouse.

Nonmaintenance Work

- Engineering, operations, and maintenance are jointly responsible for ensuring that nonmaintenance projects such as construction or equipment installation are necessary, feasible, properly engineered, and correctly funded before work commences.

- Maintenance is authorized to perform project work such as construction, modification, equipment installation, and relocation only when the maintenance work load permits. Otherwise, contractor support will be obtained subject to the current labor agreement.
- Equipment modifications will be reviewed to determine their necessity, feasibility, and correct funding prior to the work being assigned to maintenance. All such work will be reviewed by maintenance engineering before work commences.

SUMMARY

The equipment management strategy brings all plant departments together in a cohesive effort of shared responsibilities that ensure reliable equipment and a profitable operation. It transforms maintenance into a key element of plant operation fully capable of meeting the reliability demands of the more complex equipment being used in an increasing competitive global industrial environment.

CHAPTER 2

Applying the Principles of Equipment Management

PRINCIPLES

Principles are primary sources or rules of conduct. In industry, principles identify the most effective, proven way that an activity, like maintenance, should be carried out. Principles guide the manager in developing the production strategy for his plant to ensure efficiency and profitability. Principles also guide the transition from "maintenance" to "equipment management" (Figure 2-1).

Industrial maintenance management remains a complex activity. But modern production equipment designed with greater reliability and built with greater productive capacity requires a new response. It has become more complex and more difficult to maintain. Greater technical skills are required. Direct participation from personnel from all plant departments will be necessary. New strategies, revised programs, and higher quality information will be required. Reliance on time-based preventive maintenance (PM) actions of the past such as physical inspections will yield to advanced condition-monitoring technologies. Profitable industrial organizations will be required to transform older programs, procedures, controls, and organizations and adapt to new thinking. Although the basic principles of maintenance management still have application, they must also be adapted to the need for greater equipment reliability demanded by modern production equipment.

THE PRINCIPLES

Many of the proven principles of maintenance management are still valid. They now appear in a new context as the maintenance shifts to a new era of equipment management. The following paragraphs describe each of the 15 principles.

1. **Program: The maintenance program should be defined and made familiar to all plant personnel.** The maintenance program prescribes what maintenance will do, who will do what, how they will do it, and

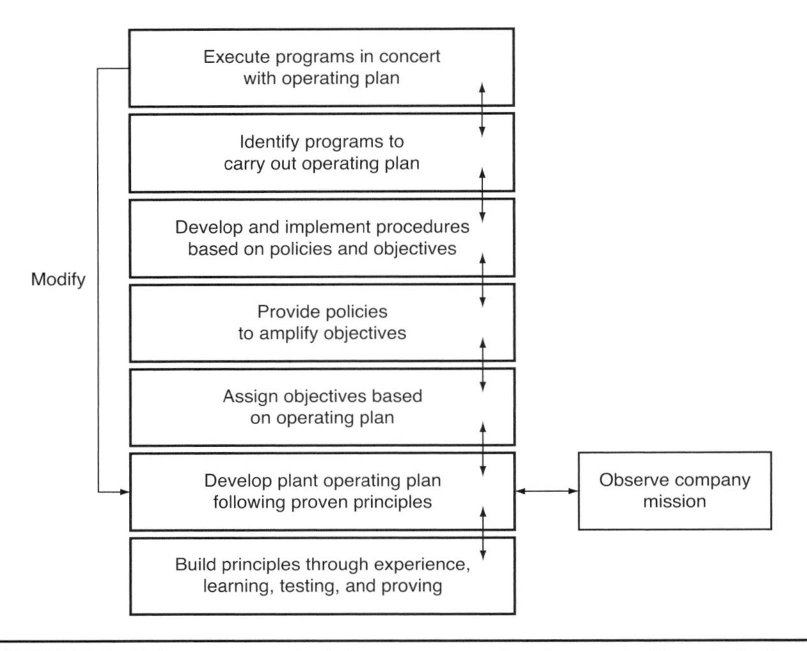

FIGURE 2-1 Using proven principles, managers develop production strategies, and provide objectives and policies. In turn, departments develop procedures based on policies and establish programs to interact with other departments.

why. The program informs maintenance personnel of necessary internal program details and interactions with other departments. In addition, program documentation ensures that all departments are made aware of how they can cooperate with and support maintenance.

2. **Work load: Maintenance should carefully identify and define the types of work they perform.** The work load is the essential work that maintenance is prepared to carry out. By carefully identifying and defining the work load, maintenance ensures its customers of their work capability and constraints. Its definition also enables maintenance to make a reasonable estimate of the proper size and craft composition of their work force.

3. **Terminology: Common words used to describe maintenance activities should be defined so that there is no confusion among maintenance personnel and people from other departments.** Uniformity in maintenance terminology avoids confusion between maintenance personnel and facilitates the correct definition of the needs of maintenance customers.

4. **Organization: Determination of the best way to organize the maintenance department should be based on the definition of the maintenance program it is intended to carry out.** There are numerous types of maintenance organizations. Each of them has specific capabilities to enable effective control of work and the efficient use of personnel. By carefully defining the maintenance program, maintenance describes what they must do, who will do what, how they will do it, and why. Based on this information, the most effective maintenance organization can be developed.
5. **Labor control: Strong emphasis should be given to the effective control of labor.** The only way that maintenance can control the cost of the work they perform is through the efficiency with which they install materials. By installing materials effectively and efficiently, quality work is performed and repairs last longer. In turn, maintenance costs are reduced by virtue of more productive use of labor and a reduced rate of material consumption. This makes the control of labor a vital maintenance function.
6. **Productivity: Maintenance should regularly measure the productivity of its work force.** Productivity is the percentage of time that maintenance personnel are at the work site with their tools performing productive work. Productivity is best when the majority of maintenance work is planned and well-organized. Effective work control by supervisors and the elimination of delays also contribute to higher productivity. Productivity is an indicator of maintenance effectiveness.
7. **Preventive maintenance: PM services should take precedence over all other work except bona fide emergencies.** Preventive maintenance drives the maintenance program. Its "detection-orientation" identifies equipment problems in sufficient time to ensure that most work can be planned. Planned work is carried out with less labor and in less elapsed downtime, yielding higher-quality, longer-lasting results. It is also the means by which maintenance can uncover serious deficiencies in time to avert emergencies. Its precedence over all other work except emergency repairs is sensible and realistic.
8. **Planning criteria: Criteria should designate work requiring detailed planning to ensure best resource utilization, timely completion, and quality work.** Major jobs warrant advanced planning because they require extensive resources and considerable scope of work. After work has been planned, it can be performed efficiently and the equipment returned to production in the shortest time. Maintenance supervisors should focus on the control of work and should not be distracted by administrative planning procedures. Criteria distinguish a supervisor's control of unplanned work from planning responsibilities.

9. **Standards: Standards should be applied in planning.** Considerable planned work is represented by jobs repeated periodically. Each repetition allows development of standard task lists, bills of materials, or tool lists. These can be applied to each repetition allowing planning to be done more consistently and effectively. In addition, because the historical labor use is also known, work load determination is facilitated.
10. **Scheduling work: All major planned work should be jointly scheduled with operations.** Scheduling should be a joint maintenance–operations activity because its objectives are to ensure that the work is done when it least interferes with operations and makes the best use of maintenance resources. Weekly scheduling meetings should be followed by daily coordination meetings to adjust the schedule in the event of unforeseen delays. Compliance with approved work schedules should be measured to ensure that work is completed as prescribed and interdepartmental cooperation is verified.
11. **Priority-setting: A priority-setting procedure should be utilized so that maintenance and operations can effectively determine the relative priority of work and the time within which the work should be completed.** The most effective priority-setting schemes would allow operations to communicate the criticality of the equipment affected while maintenance would be able to indicate the importance of the work to ensure reliability of equipment. In addition, operations, as the customer, should be given a reasonable expectation of the time when the work should be completed.
12. **Information: Information systems should be utilized to ensure efficient requesting, identification, control, and measurement of work.** The information and work order system should provide procedures for requesting, identifying, controlling, and measuring work. Any department requiring the control of work should utilize it. Field data should be initiated by the most reliable source to ensure it is timely, complete, and accurate. The information system should convert field data into accurate information so that key personnel can make well-informed decisions.
13. **Material control: Maintenance should carefully follow material control procedures.** Maintenance is the plant's largest consumer of materials. It depends on support from warehousing and purchasing to provide repair materials. These organizations have procedures to ensure effective support of maintenance. Maintenance personnel should follow procedures and inform material control personnel of maintenance program details to ensure mutual cooperation.

14. **Nonmaintenance: Nonmaintenance work such as equipment installation should be assessed to ensure it is necessary, feasible, and correctly funded before being assigned to maintenance.** Maintenance organizations may be called on to perform nonmaintenance work such as equipment installation or modification, as the same craft skills are used. If there are no maintenance personnel regularly dedicated to this work, performing it in excess could undermine the ability of maintenance to carry out basic maintenance. Nonmaintenance jobs are often complex and time-consuming. They should be properly engineered before being assigned to maintenance. Often, this work is capitalized rather than expensed. Maintenance should verify the funding source to avoid misapplication of the maintenance budget.
15. **Maintenance: Maintenance should evaluate all factors that influence its performance and act with other departments to ensure improvement.** Numerous factors affect maintenance performance. Maintenance, for example, depends on other departments for material control support but does not control these departments. It is dependent on these departments to provide quality service. If the service is not adequate, maintenance performance is adversely affected. Therefore, when maintenance evaluates its performance, it should also evaluate material control support or cooperation from operations. Similarly, they should invite other departments to examine mutual working relationships.

SUMMARY

Using proven principles as a guide, the maintenance program can be carefully developed, the proper organization established, and procedures set in place to ensure effective maintenance. These principles are also used to propel the maintenance program into an effective equipment management activity. Emphasis is given to the need for mutually supporting interaction among departments and the introduction of modern reliability strategies plus the effective use of information. These principles provide guidelines for this task.

CHAPTER 3

Developing the Equipment Management Program

THE STRATEGY IN PERSPECTIVE

The principles of the equipment management strategy guide the development of the production strategy, the plant manager's plan for ensuring the profitability of a plant. In turn, the production strategy establishes the basis for an effective maintenance program. Together, the principles, production strategy, and maintenance program constitute the equipment management program. The maintenance program now guides the collective efforts of all plant departments, enabling them to deliver reliable equipment consistently. The plant's production strategy is now met. The principles of equipment management have been applied and the company's mission satisfied. In the process, the profitability of the plant is better ensured.

EXAMINING THE MAINTENANCE PROGRAM

The maintenance program spells out how production will *request* work and how maintenance will *identify* it through their preventive maintenance (PM) efforts. The program describes how new work will be *classified* to ensure the best response. It will define how maintenance will *plan* selected major jobs so they can be performed effectively. It will also prescribe how operations and maintenance jointly *schedule* planned work to ensure it is done at a time that interferes least with operations and makes the best use of maintenance resources. The program will prescribe how supervisors *assign* work and *control* it with effective supervision. The program effectively *measures* completed work against targets like costs and timely completion. Next, the program will assess overall accomplishments against performance standards or budgets. The program will *evaluate* everything that affects maintenance performance to identify and prioritize improvement needs. Evaluations will be part of a plantwide continuous improvement effort. The development, documentation, publication, and effective utilization of the maintenance program will be the most essential element of a successful equipment management effort. Advanced strategies such as total productive maintenance or reliability centered maintenance are then integrated into the maintenance program to meet the special needs of enhanced equipment reliability.

GETTING STARTED

Effective definition of the maintenance program avoids confusion and uncertainty. Operations will know how best to cooperate with maintenance. Staff departments, like warehousing, won't try to second-guess what maintenance wants. By stating, "I want a logical, well-defined program from maintenance that fits our production strategy, and I want all plant personnel to understand it, support it, and make it work," plant managers will appreciate the effective interactions that result.

By defining the maintenance program, existing program elements are confirmed and new program elements can be developed more efficiently. Yet the process of defining the program is beneficial. Greater participation in change is encouraged, and commitments are made with better understanding of program objectives. Invariably, the work force will be well-informed and more able to help. Program definition also has great potential value in improving interdepartmental cooperation. The program clarifies the specifics of interactions among plant departments. Warehousing and purchasing, for example, will specify procedures to obtain materials, whereas maintenance will incorporate these procedures in their program to better ensure solid, interdepartmental cooperation.

PROGRAM DEFINITION

Full cooperation among departments requires clear, logical management guidance. Therefore, program definition begins with the plant manager. First, he ensures that his production strategy supports the company's mission. Next, he assigns objectives to each department specifying how he wishes them to work together. Then he provides policies to guide department interactions. From his policies, departments develop mutually supporting procedures. Internal maintenance procedures (e.g., scheduling PM services) and interdepartmental procedures (e.g., how maintenance would obtain stock materials) are woven together to ensure the practicality of the program. Education about the program should immediately follow program definition. As the program is explained, questions and recommendations should be encouraged. All maintenance and operations personnel as well as staff departments should be included. Plant managers should observe department discussions and reassert the need for interdepartmental cooperation.

PROGRAM DEFINITION TECHNIQUES

The most effective technique for documenting the program is a schematic diagram that depicts the interaction between individuals of participating departments. The schematic pinpoints "you" and "me" and describes what "we" must do, how we will do it, and the results we should achieve. This "personal" explanation helps to bind people to the program while facilitating the educational process and encouraging participation. The initial schematic is accompanied by a proposed legend to aid in understanding the step-by-step process.

Initial program development should not be dictated from the top down, as it will appear that maintenance managers are not open to sensible suggestions. Instead, the initial program should welcome the participation of as many personnel as possible. This approach not only encourages constructive suggestions, many with great merit, but it also creates a sense of ownership in the final program and a greater commitment to ensure that all personnel will work harder to make the program successful. In addition, the more that personnel understand the program and its objectives, the more they will watch its implementation and be better prepared to step in and make adjustments should minor problems arise that need immediate attention.

Preventive Maintenance Procedures

Figure 3-1 and its accompanying legend illustrate the interactions within maintenance and between maintenance and operations as PM services are scheduled, carried out, and results acted on. Such a diagram facilitates the explanation of procedures as the program is launched and provides guidance after the program is in effect. When two operating departments interact, as with Figure 3–1 that depicts how operations and maintenance cooperate to ensure the success of the PM effort, the outside department (operations) is given opportunity to share their concerns and often will make many constructive suggestions. When they see that their recommendations have been incorporated into the program, they will invariably work harder to ensure the success of the activity they now fully understand and better appreciate how it will make their efforts more effective.

The narrative in the legend should be kept as brief as possible. The basic message simply needs to be conveyed. Then, during the explanation of the program, personnel can indicate whether the explanation needs to be expanded or clarified.

Multiple Department Procedures

Multiple department interactions will require especially careful coordination for them to work effectively. Therefore, significant details of each department must be well known to all departments. If, for example, a senior maintenance

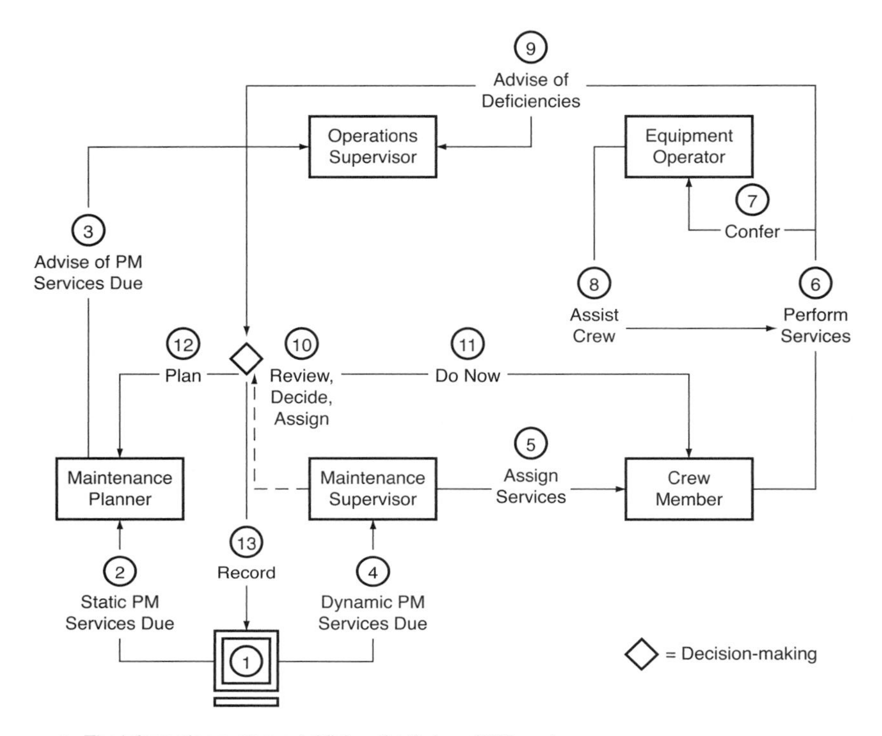

1. The information system establishes the timing of PM services.
2. Services on equipment due are either static (require shutdown) or dynamic services (performed while equipment is running).
3. Static services are integrated into the weekly schedule and operations supervisors are advised of the approved, scheduled shutdown times.
4. Dynamic services are performed at the discretion of the maintenance supervisor.
5. The maintenance supervisor assigns services to individual crew members.
6. Services are performed by crew members.
7. Crew members confer with operators to learn about actual equipment condition.
8. Operators assist according to checklist instructions.
9. Operations supervisors are advised of new deficiencies by the crew member.
10. Deficiencies are then reviewed by the maintenance supervisor and the crew member and converted into work as follows:
11. Emergency repairs—Supervisor assigns at first opportunity;
12. Work to be planned—Supervisor forwards to planner;
13. Unscheduled repairs—Crew member makes an entry in the work order system as new work.

FIGURE 3-1 Preventive maintenance procedures

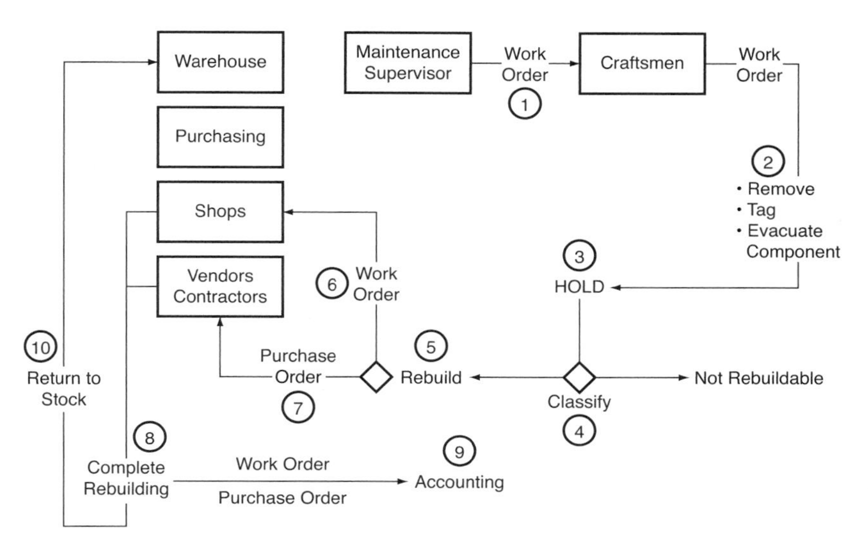

1. The maintenance supervisor assigns the work order to remove the component.
2. In turn, the craftsman removes, tags, and evacuates the component to the designated area.
3. In the holding area, the component is logged in by warehouse personnel and maintenance is notified of its receipt.
4. The component is classified to determine whether it is able to be rebuilt. If not, it is scrapped. If a rebuild is possible,
5. It is sent to either plant shops or vendor shops for rebuilding.
6. If sent to internal shops, it is accompanied by a work order provided by the warehouse.
7. If a vendor will perform the rebuild, a purchase order is provided via the purchasing department.
8. When the rebuild is complete, accounting
9. Receives the shop labor and material charges or the vendor's invoice.
10. Then the rebuilt component(s) is returned to stock following suitable quality control checks prior to reissue for use by maintenance on future jobs.

FIGURE 3-2 Interactions among multiple departments

supervisor and a planner were to unilaterally devise a multiple-department procedure, the essential details of warehousing or purchasing operations would probably be incorrect, oversimplified, and ultimately ignored. Therefore, every department involved should participate in program development.

Figure 3-2 illustrates the interaction of several departments as they go about the common tasks of component rebuilding.

Shop Utilization Procedures

The coordination of shop services with field planners and line supervisors is illustrated in Figure 3-3.

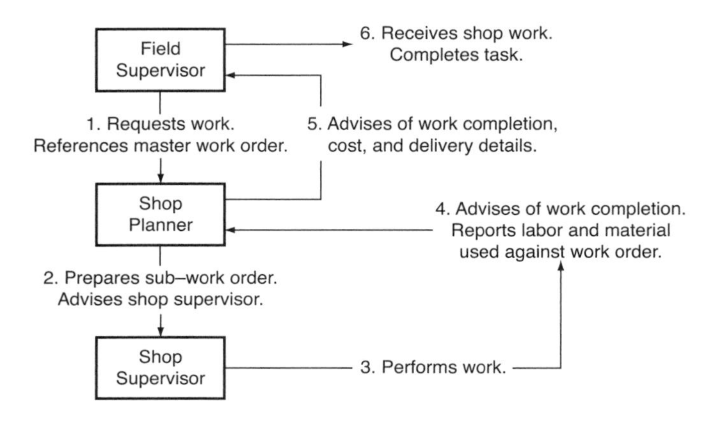

1. The maintenance field supervisor requests shop work from the shop planner and references the master work order.
2. The shop planner prepares a sub–work order and advises the shop supervisor of work required.
3. The shop supervisor performs the work.
4. The shop supervisor advises the shop planner of work completion and reports labor and material used against the work order.
5. The shop planner advises the field supervisor of work completion, costs, and delivery details.
6. The field supervisor receives the shop work and completes the task.

FIGURE 3-3 Coordination of shop services

Work Order and Accounting System Interface

The information system is the communications network for the equipment management program. Its utilization in supporting the program should be incorporated into program definition by illustrating the information system–accounting system interaction.

One of the most significant and damaging aspects of effective work order utilization is the failure of key maintenance personnel to understand how the accounting system functions. Many will simply transcribe accounting numbers that are "safe" (i.e., will not get them in trouble) but are often incorrect, and sometime laughable, in the eyes of accounting personnel. Similarly, accounting personnel are mostly interested in numbers and costs, and often ignore information that is vital to maintenance, such as the narrative describing repair history. When these two voids are not acknowledged and corrected, the unfortunate result is often an information system that meets accounting needs but is impractical from a maintenance point of view. Therefore, mutual discussion and education between maintenance and accounting personnel during program development is vital to program success.

Figure 3-4 illustrates the work order and accounting system interface.

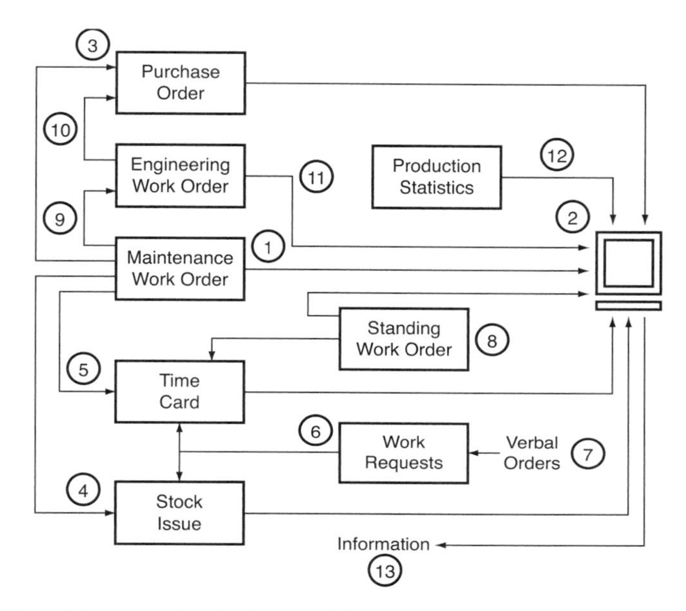

1. The maintenance work order is a control document used to isolate the cost and performance of a single major, planned job. Each maintenance work order is assigned a unique work order number to relate it to the specific job as well as to the equipment type, number, and component.
2. The work order is opened in the database, making it an official accounting document ready to receive labor and material data input.
3. The planner then establishes a link with purchasing to obtain direct charge materials by placing the work order number on the purchase order(s).
4. Next, stock materials are ordered and the work order number is placed on all stock issue documents.
5. As work is performed or completed, labor data are recorded on the time card or other procedures for reporting labor, and the labor used is associated with the proper work order number.
6. Most systems have a means of providing work requests in which case-related labor and material data flow into the database through existing accounting or system procedures.
7. Similarly, verbal orders may be used with comparable methods of reporting data.
8. Standing work orders identifying routine, repetitive tasks like training or shop cleanup are opened in the database to accumulate costs for an entire year.
9. Should maintenance perform nonmaintenance work, like new equipment installation, an engineering work order is established. Actual project work may be carried out by either maintenance or a contractor or both. Project work performed by maintenance is still controlled by the work order but the work order number must be associated with the engineering work order number.
10. Direct charge materials are associated with the project by linking the engineering work order number with the purchase order number. If, however, the materials are used by maintenance to complete their part of the project, the purchase order and work order numbers would be linked instead.
11. Now the engineering work order is opened in the database to control the project.
12. Production statistics are often added to allow comparisons of, for example, cost per unit of product to be made.
13. Finally, all of these elements are brought together in the database to produce information such as cost, repair history, job status, etc.

FIGURE 3-4 Interface between work order and accounting system

Legend for Equipment Management Program Schematic

The equipment management program shown in Figure 3-5 is typical of the detail that a large plant would consider in the initial development of their maintenance program. The following legend would accompany such a program. It is recommended as a possible model for the development of a practical, realistic maintenance program. It illustrates an effective way to explain the maintenance program to personnel from all departments. As you read the diagram, follow the numerical sequence, noting the captions on the diagram and reading the corresponding explanation in this legend.

1. Operators perform periodic equipment checks to establish whether equipment is running properly. While the design of modern equipment often provides monitoring systems or protected systems with alarms and signals, the additional personal attention of operators often finds the unusual problems that might be missed. Moreover, these checks exhibit the interest of the operator in the equipment, a fact very much appreciated by the maintenance craftsmen. Operators submit reports of equipment condition to operating supervisors.
2. Operations supervisors, in turn, review the operators' reports as a potential source of future work.
3. Operations supervisors then determine whether new work is required and specify the unit, the problem, and the urgency of the job, and then identify themselves as work requesters using the standard work request procedure.
4. If the work required is an emergency, the operations supervisor could make a verbal request to expedite repair.
5. Upon receipt of work requests, the maintenance supervisor initially classifies the work. If the work requested meets the criteria for being planned, it would be forwarded to the planner. In the case of emergency repairs, the maintenance supervisor would assign work immediately. Work requests for an emergency would be created to facilitate reporting of resource use such as labor and materials by maintenance crew members.
6. The maintenance supervisor would advise the planner of all work that requires planning based on his assessment of the planning criteria. As necessary, he would advise the planner of any special requirements that the work entails.
7. Maintenance planners utilize the information system to determine which PM services are due the following week. The information system identifies PM services due based on fixed time intervals such as weeks or variable intervals like operating hours. Ordinarily, the PM module of the information system is updated once a week, and at that time services

due the following week are identified and placed on the preliminary maintenance plan. Some PM services are dynamic and can be carried out while the equipment is running. These PM services are assigned directly to the maintenance supervisor so he can fit them into the week's work at the first opportunity. Other PM services due are static services, requiring that the equipment be shut down in order to perform the work. These services are handled by the planner. They are fitted into the weekly plan along with major jobs.

8. The information system should have the capability of forecasting the approximate timing of the replacement of major components, such as engines and drive motors. In addition, the approximate timing of major unique jobs such as overhauls is also forecasted. These forecasted actions are normally displayed week by week so that the planner may query the information system and determine what actions are due the following week. After these are identified, the planner determines the actual condition of the equipment, and, after conferring with maintenance supervisors, may either place the activity on the proposed plan for the following week or defer the work based on current repair history and the actual operating condition of the equipment.
9. After the planner has assembled all of the potential work that would be attempted for the following week, she roughs out a preliminary plan to include major component replacements, unique major jobs such as overhauls, and static PM services. The plan does not become a schedule until it has been presented to operations personnel and they concur that the plan should be carried out.
10. In carrying out the planning steps, planners address the following issues:
    - Identify the work required
    - Investigate in the field to determine exact needs
    - Seek advice from the supervisor who will do the work
    - Determine whether any standards apply
    - Confirm the job scope
    - Make a job plan
    - Set up the preliminary work order
    - Determine resources that will be needed
    - Establish the labor requirements by craft
    - Identify materials needed and the sources
    - Specify any shop work
    - Estimate job cost
    - Establish the job priority with operations

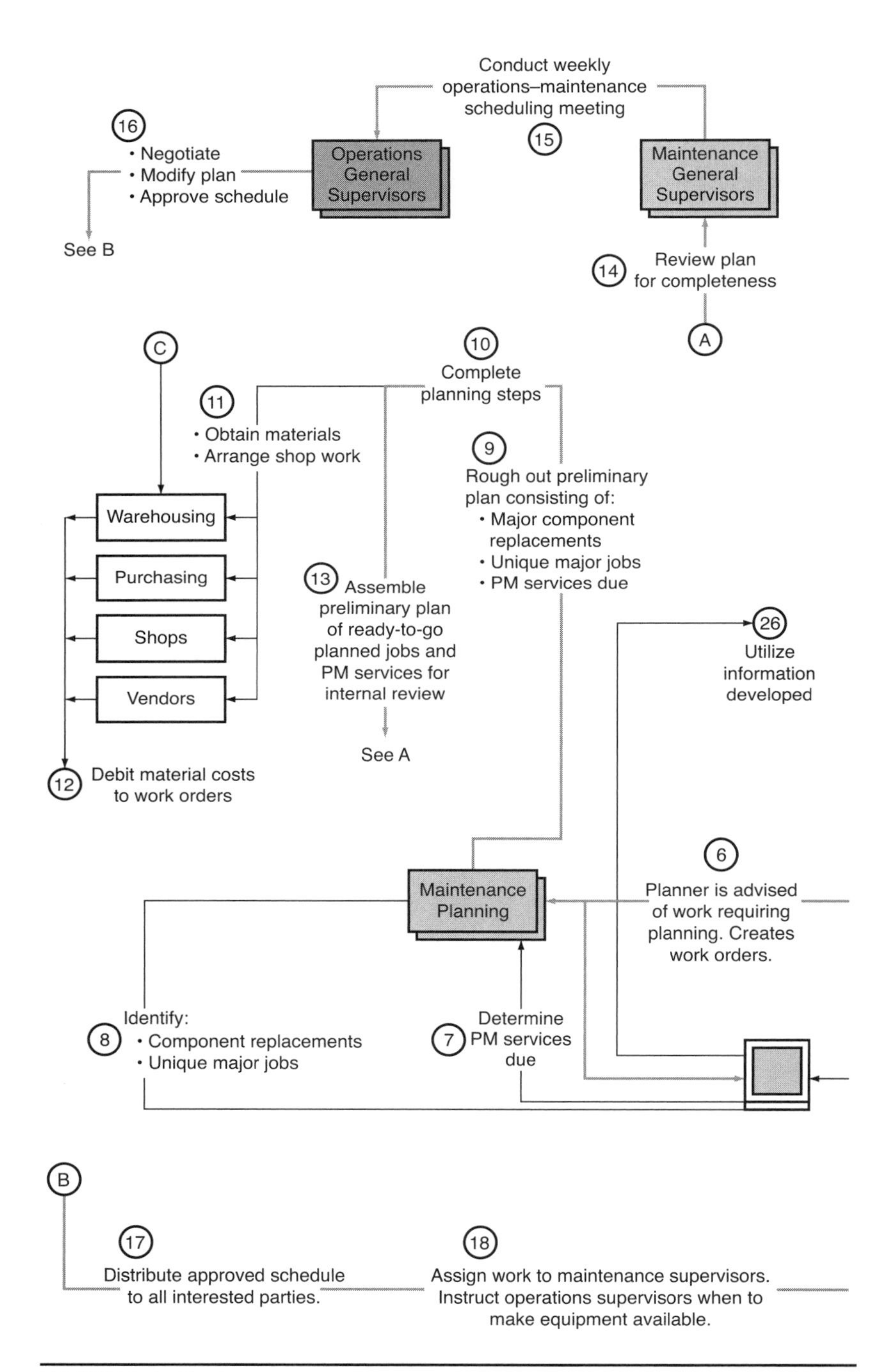

FIGURE 3-5 Equipment management program schematic

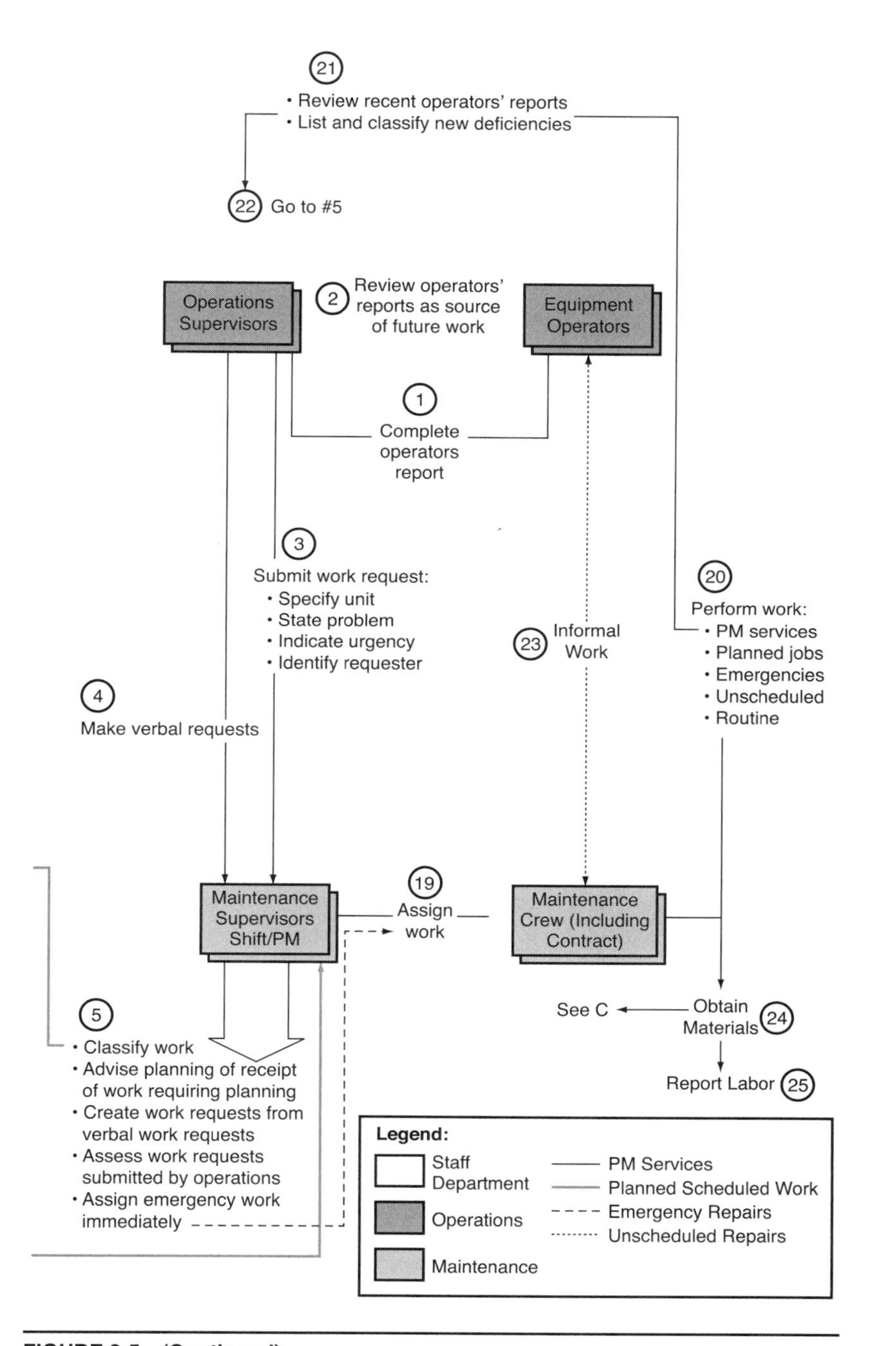

FIGURE 3-5 (Continued)

- Get the job approved: cost and preliminary timing
- Identify the possible future time for doing the work
- Open the work order, making it an "official" accounting document
- Order materials through purchasing
- Reserve the stocked parts
- Order the shop work
- Await the receipt of materials and completion of shop work
- As materials are ready, confer with operations on job timing
- Develop a preliminary weekly plan
- Review plan internally before the scheduling meeting with operations

Information on the repeated conduct of work, such as the replacement of major components, is used to develop standards for such jobs. These standards include task lists, bills of materials, and tool lists. Timing of component replacements is based on the historic life-span of the components. Many organizations have developed forecasting procedures to help identify the timing of components replacement.

With the advent of more sophisticated condition-monitoring techniques, forecasting has become a dependable source of information for maintenance actions. It simplifies planning by providing proven standards for repetitive tasks. As unique major jobs are being planned, the maintenance engineer is called on to establish the job scope. The job scope specifies the work necessary to restore the equipment to an as-designed condition. The maintenance engineer prescribes steps that will ensure the reliability of equipment and the application of techniques that ensure its maintainability. With the requirement for using more sophisticated condition-monitoring techniques to secure the increased reliability, this role is more important than ever. Predictive techniques must be used expertly. In addition, maintenance engineers will be scrutinizing the entire program. They will verify the application of PM techniques, the use of modern technology, and the effectiveness of information use. The maintenance engineer can contribute significantly to the successful implementation of new technologies required for equipment management programs.

11. Planners reserve stock materials required for major jobs to be carried out for the following week. In addition, they instruct the purchasing agent to procure direct charge materials. Planners use the information system's purchase order tracking system to monitor the status of these purchase orders. They also order the shop support, confer with the shop planner, and provide sub–work orders as required. They monitor shop work progress using the information system's work order status report.

12. The cost of materials is debited to the appropriate work orders so the total cost of each job along with the cost of labor can be established. When the work is completed, the total cost of each job can be determined.
13. As materials and shop work become available, the planner assembles a preliminary plan of all candidate work orders and discusses them internally with maintenance supervisors to ensure that all are feasible and ready to go. See A.
14. The preliminary maintenance plan for the following week is reviewed by maintenance supervisors to determine whether it is feasible and complete or if certain jobs might be deferred or canceled.
15. The approved maintenance plan is then presented to operations by the maintenance supervisor at the weekly operations–maintenance scheduling meeting. It is important that the supervisor present the plan to operations because he is ultimately responsible for carrying it out. The planner provides a supporting role during the scheduling meeting in the event that job priorities must be changed or resources rearranged to be able to carry out the schedule effectively. The weekly scheduling meeting with operations is the opportunity for maintenance to present its plan for accomplishing major jobs and key static PM services for the following week. Based on the production plan, operations will consider the needs for equipment and the "mix" they must have against the requirement to perform maintenance at the recommended times. Similarly, the shutdown of fixed equipment, lines, or areas is considered. Necessary negotiations are carried out by the principals with the coordinating planner acting as an advisor. Scheduling meetings should be chaired by the decision makers who will be responsible for making equipment available and accomplishing the work. Given the accountability demanded of all departments, purchasing agents and warehouse supervisors should attend scheduling meetings to ensure that their responsibilities are being met. The scheduling meeting should conclude with an approved schedule. Thereafter, operations should make equipment available according to the schedule, and maintenance would make its resources available to perform the work. The objective is to accomplish the work with the least interruption to operations and the best use of maintenance resources.

During the week when scheduled jobs are being done, daily coordination meetings between operations and maintenance are held to make adjustments in the schedule in the event of any delays.

Many plant managers use the weekly scheduling meeting as an opportunity to assess progress against their overall improvement goals such as

- PM program compliance
- Reduction in backlog
- Compliance with last week's schedule
- Reduction in overtime
- Reduction in emergency repairs
- On-site parts delivery performance by the warehouse

16. As a result of the negotiations carried out between maintenance and operations during a scheduling meeting, the plan may be modified to better meet the needs of operations and to utilize maintenance resources more effectively. The scheduling meeting is followed by daily coordination meetings between operations and maintenance to adjust the schedule in the event of delays. See B.
17. The approved schedule is now distributed to all interested parties.
18. Maintenance supervisors are now assigned the work specified on the approved schedule. This step suggests that each supervisor has the opportunity to coordinate directly with purchasing, shops, or the warehouse for on-site material delivery. In addition, transportation and service departments should help coordinate their support. The supporting planner might be asked to assist in coordinating for the supervisor who will be on the jobsite controlling the work. Supervisors assign work on the schedule to individual maintenance personnel. They provide appropriate work orders and necessary instructions for carrying out the work. In addition, operating supervisors are advised when they should make equipment available so that maintenance work can be carried out.
19. Using the weekly schedule as a guide, maintenance supervisors give work assignments to maintenance crew members on a day-to-day basis. As necessary, supervisors consult the information system to determine whether there are any pending small jobs on the same equipment scheduled for shutdown and whether these smaller jobs might be fitted in.
20. Maintenance crew members perform PM services on planned jobs, respond to emergencies, perform unscheduled repairs, and carry out routine services.
21. Craftsman and operators confer during the inspection of equipment by maintenance crew members. Operators can explain unusual equipment behavior while craftsmen can show operators simple, helpful checks or adjustments. With mobile equipment, operators are often not present. Therefore, comments are provided based on operator check sheets. As a result of personal inspection, operator input, and results of the operator's checks, the craftsman can make an accurate, complete list of equipment deficiencies and their relative seriousness. These deficiencies should be

identified to the equipment component level. Generally, there are three types of deficiencies:

- Emergency repairs. Deficiencies found just in time to avert an injury, equipment failure, or serious loss of product or production time are usually emergency repairs. These jobs should be acted on as soon as possible and, in most instances, key people advised in the event that other actions, like the use of alternate equipment, are necessary.
- Unscheduled minor repairs. The most numerous deficiencies found are those smaller repairs that do not have to be done immediately. These are unscheduled jobs. But because they can develop into more serious problems, they must be recorded and fitted in at the first opportunity. (Maintenance supervisors should instruct craftsmen to always finish the whole PM service before making any lengthy repairs. This procedure should be followed because the PM service must be performed within a specific time limit. If too many small, incidental repairs are made "as you go," there is a good chance that the total service might not be completed. Thus, there is a danger that problems not found are more serious than those repaired. Conversely, it is a good idea to add a few minutes to the specified time to make simple adjustments that require only tools or limited materials just to save a return trip.)
- Major repairs. A limited number of deficiencies will indicate a need to make significant repairs. These are potentially planned jobs. But if PM services are conducted faithfully, these jobs will always be found far enough in advance to be able to plan the work. Craftsmen should make a special effort to establish the exact nature of the problem so it can be accurately described to the maintenance supervisor and, in turn, to the planner.

Craftsmen should advise production supervisors directly on the deficiencies they have found and how serious they are. This thoughtful act not only does the operations supervisor a great favor in giving her a current condition report, but it lets her know that the PM program is alive and well.

Moreover, when she learns that each deficiency found will be taken care of, she sees a solid partnership developing with maintenance people. She is also delighted to know that her operators have contributed with their checks. This action spreads good will and cooperation.

22. Go to #5. Craftsmen get together with the maintenance supervisor and together they decide on the actions to be taken on the deficiencies reported. Craftsmen supply firsthand information on actual equipment condition, and supervisors provide knowledge of the bigger picture of what to do, when to do it, and how. Specific decisions are now made on

each type of deficiency relative to equipment condition, the urgency of corrective actions, the current plant situation, and the availability of maintenance resources. First, the emergency repairs are acted on. Generally, verbal orders are used to assign work to available craftsmen with followup work orders after the repairs are attended to. Next, those deficiencies requiring planning (#6) are discussed with the planner. In this instance, the prudent maintenance organization will already have provided sensible criteria of what work is to be planned. Thus, the planner, supervisor, and craftsman are in full accord. The craftsman performing the PM services should enter all of the unscheduled new jobs into the work order system. Remember that the craftsmen performing the work are the best source of information. Always use them.

23. As maintenance services are performed, craftsmen may also be called on while near or on the equipment, as with mobile equipment, to provide informal help. Generally, supervisors should have provided criteria to the craftsmen to determine whether any of the informal repairs made should be entered into repair history.
24. In the process of doing work, maintenance crew members may require additional stock materials. See C.
25. Upon completion of work, maintenance crew members report the labor used. This data is subsequently verified by the maintenance supervisor. Labor and material data recorded on work orders, along with other information such as failure codes, is processed by the information system to provide information reports for managing maintenance, such as labor utilization, cost reports, repair history, work order status, and backlog.

SUMMARY

Successful implementation of the equipment management program will require plant personnel to think about how they will perform maintenance. They will also consider how to best use modern management techniques, new technologies, and information to achieve greater equipment reliability and full productive capacity. The act of defining the maintenance program is an opportunity to realign responsibilities to apply the new program elements. In the process of defining the program, greater participation is encouraged and commitments are secured as a result of better understanding.

CHAPTER 4

Leadership in Equipment Management

TRANSITION TO EQUIPMENT MANAGEMENT

When an industry makes the decision to transform their maintenance function into a plantwide equipment management program, they must convert maintenance thinking to equipment management thinking from the board room to the worker level. Thus, the following term,

> Maintenance—The repair and upkeep of existing equipment, facilities, buildings, or areas in accordance with current design specifications to keep them in a safe, effective condition while meeting their intended purposes.

becomes . . .

> Equipment management—A fully coordinated, mutually supporting effort of every plant department and individual to achieve maximum reliability and productive capacity of critical equipment throughout its entire life cycle.

The success of this transition into this much broader, strategic view of what used to be called *maintenance* will require firm, thoughtful management leadership.

Industrial managers must cause this new thinking to be absorbed and new responsibilities accepted. They must establish a plant environment that convinces all personnel of the urgency of approaching maintenance with a new, more effective overall strategy.

A strategy such as equipment management should be introduced at the top of the organization. Leaders, all down the line, should be guided by a single corporate objective: to be provided with solid guidance and incentives to transform their organizations. Therefore, starting with executives, there must be a full understanding of what can be accomplished and how.

EXECUTIVE LEVEL

Executives will have carefully examined the marketplace to determine the cost levels at which their products can effectively compete. After determining profit goals, they will establish a margin between current operating costs and the competitive market costs they must meet. These factors will influence their strategies to meet the lower operating costs. They may consider downsizing, closing unprofitable plants, or merging with other companies to reduce the cost of goods supplied to them or acquire products that complement and enhance their own marketplace appeal.

Individual plants must be lucrative to ensure a profitable corporation. Therefore, realistic and workable programs like equipment management must make their primary marks at the plant level. Yet individual plants cannot embark on strategies that are not sanctioned by their total corporation. In the case of equipment management, traditional responsibilities of given departments will be changed and broadened. For example, in operations that have relied on traditional maintenance, purchasing agents rarely attended the weekly operations–maintenance scheduling meeting. With equipment management, they are required to provide an up-to-date status of all pending materials procurement, explain the quality control measures being applied to vendors rebuilding components, or explain why a shutdown did not start on time when critical materials failed to arrive. Moreover, the purchasing agent must then identify the steps to be taken by his department to ensure improved future performance. Most plant departments have a corporate counterpart. Therefore, the director of purchasing, for instance, must agree to and sanction alterations of plant purchasing department activities and responsibilities. Thus, the successful plant-level implementation of the equipment management strategy requires corporate-level sanction, commitment, and informed guidance to plant managers and department managers. Corporate executives responsible for overseeing multiple plant operations must start by examining the potential for improvement offered by the equipment management strategy. The strategy will focus on preserving equipment functions rather than the equipment itself. It will

- Place a premium on avoiding the far-reaching consequences of failure rather than just the equipment failures themselves;
- Preclude the damaging consequences that reach beyond the failure of equipment (injuries, lost production, equipment damage, environmental damage, etc.);
- Require application of modern maintenance engineering (latest technology, advanced condition monitoring, quality information, and pertinent management techniques);
- End "seat of the pants" maintenance;

- Acknowledge that maintenance personnel must change, adapt, and be educated in new thinking; and
- Specify that every department and every individual have a pertinent, direct role and specific responsibility for equipment management.

When implemented, these actions can ensure

- Greater equipment reliability,
- Less downtime,
- Reduction in maintenance costs,
- Realization of the full productive capacity built into equipment,
- Maintenance can be done with fewer people,
- Maintenance will be required less often,
- Lower total operating costs, and
- Greater margins for profits.

Executives should ensure that the fundamentals of maintenance management are in place and fully functional at each plant. They should then verify each plant's capability to effectively carry out the equipment management program, regardless of how the maintenance function is controlled or organized. Senior managers should also verify that there is an effective information system and it supplies good information to the right personnel, who can then make correct decisions and take proper actions. Executives should ensure that plants have provided clear objectives to each of their departments as well as policies to guide development of the interdepartmental procedures that will comprise the equipment management program. All of these points are then summarized in a thoughtful corporate mission that stresses corporate goals and characterizes actions leading to the goals:

> We wish to be acknowledged by our customers as a world leader in the efficient manufacture of quality products. We will accomplish this goal with the successful implementation of strategies that make the most effective use of our human and physical resources.

Successful corporate implementation of the equipment management strategy requires that executives play an active, supportive role in helping plants. For example, a vice president of operations should continue to require a report on production achievement. But she must now also ask whether the plant has achieved contributing objectives like a regular 85% compliance with the preventive maintenance (PM) program.

Therefore, as the equipment management program is given serious consideration, executives must verify that plants have mastered the fundamentals of effective maintenance. Then they can start changing the way plants think about maintenance by establishing different, broader roles at every level of the organization.

Senior executives play a key role in establishing a favorable environment for the successful implementation of the equipment management strategy at the plant level. They should tactfully verify the achievement of maintenance fundamentals at each plant and then coordinate the corporate-level staff support during the implementation phase.

PLANT MANAGER

The most important role a plant manager can play in the transition from traditional maintenance to effective equipment management is to ensure that the fundamentals of maintenance are well-established and fully functional. These fundamentals of maintenance management include his provision of clear objectives for maintenance, plus policies that guide the interaction of departments as they cooperate in carrying out the maintenance function. Further, he should verify that the maintenance organization is properly organized with effective supervision and all work is well-controlled.

The plant manager would confirm these matters by ensuring that the maintenance program is working effectively, evidenced by high productivity of workers, maximum planned work, minimum equipment downtime, and responsive fiscal control of activities. He should not rely strictly on summary reports but should visit plant work sites to see for himself how the actual work is being done and whether operations personnel are satisfied with the results.

MAINTENANCE MANAGER

The maintenance manager sees "help" on the horizon, in the form of accountability and responsibility for actions that he alone was totally responsible for. For example, operations used to send projects directly to maintenance with little thought that the projects needed capital funding and engineering. The operations expectation was that maintenance would somehow get them done. Now, all nonmaintenance work must go to plant engineering to determine whether the project is feasible, necessary, and properly funded. Should it meet these criteria, it would be done, but not necessarily by maintenance. Contractors might be used.

Thus, the maintenance manager sees a whole range of former "political" issues being solved by the new program. However, this elation might be short-lived since he must deal now with the more mundane issues. He must now realistically deal with the maintenance supervisors who have been busy ignoring the new "million dollar" information system. He must now face the issue of establishing the maintenance engineering function to cope with the coming "wave" of new condition-monitoring technology. On a human side, he must start working with the hourly work force (or bargaining unit) on the prospect that the new strategy will require that maintenance, in the future, be done with fewer people (better productivity as well as a work force reduction) and that it be carried out less often (a smarter program with modern technology is required). In addition, if a self-directed team is planned for the future, he may not have a job nor will any of his supervisors. He must ensure that the hourly work force is up to the task. He must start their education and explore new roles for himself as well as his ex-supervisors. Planners in a team environment can't be clerks anymore. They must plan for new craftsmen clients. Thus, the maintenance manager should read the plant manager's "conceptualization" very carefully to establish an approach to his new "world." Then he must take steps to prepare those who will help him: craftsmen, supervisors, planners, and so on.

OPERATIONS MANAGER

The most likely organizational trend of an operations organization in the future will be the formation of business units, in which operations is responsible for both operations and maintenance. Thus, unexplained "production loss" can no longer be pinned on a hapless maintenance organization as some mysterious "mechanical failure." But the operations managers must now start to deal with reality if there is to be a new business unit approach. Their supervisors, for example, long used to the behavior of meeting production targets first and asking questions later are unlikely to simply assume total maintenance responsibility. Instead, they will rely on the craftsmen to see to the details of maintenance. In turn, these craftsmen, used to seeing major components "magically" arrive when needed, may not be able to produce the same magic. Soon, the well-intended value of the business unit will start to unravel. Thus, the typical operations manager should carefully read the plant manager's conceptualization and learn of the new responsibilities she will almost certainly acquire. From that instant on, she should seek the solace and help of her good friend, the maintenance manager.

KEY STAFF MANAGERS

Principal departments such as accounting, data processing, warehousing, and purchasing will all be required to make adjustments to accommodate their new responsibilities for the equipment management strategy. The major component replacement forecast is a good example of how their total, and integrated, responsibilities will be affected. The following experience (case study) comes to mind:

> A major drive assembly, not rebuilt to quality standards, was installed only to fail prematurely. Following grumbling by operations, they wrote off the downtime to some mysterious malady. Maintenance, unsure of whether the failure was due to putting it in backwards during an emergency repair, replaced the component without "ceremony." The damaged component, untagged, was found 6 weeks later by the

warehouse driver who took it to the warehouse classification point where it should have gone on its "classification, failure investigation, and rebuild journey" much sooner. Meanwhile, an identical component was ordered only to find a "stock out." Recriminations followed and it was learned that the whole procedure for rebuilding components had failed. More details were uncovered, including the fact that no actual written procedure existed, but most who were involved had various "folklore-based" notions of what was supposed to happen.

In equipment management, procedures like this must be documented in the maintenance program and every person involved in each step made fully conversant with not only his or her role but everyone else's as well. Therefore, as maintenance personnel remove a damaged component, they must tag it with specified data and remove it to a pickup area. In turn, the warehouse would classify the component and determine the cause of failure. From that point, local shops or designated vendors would accomplish the rebuild guided by quality standards established and verified by purchasing. After the quality standards are verified, the component is restocked, assigned a new unit value, and added to warehouse inventory. Accounting measures follow up with invoice payment to vendors or identification of work order costs associated with in-house shop rebuilding. The procedure continues with maintenance tracking each serial-numbered component in their repair history. The purchasing agent reviews the maintenance forecast for information on the "rate of consumption" of components and immediate future needs. Then, on-hand warehouse stock of these items are routinely checked by all parties. Finally, the plant manager, through his "continuous improvement" effort, periodically verifies that such important interdepartmental procedures function according to his conceptualization. Once satisfied, his "expectations" become guidelines and then policies.

In the previous case study, the gaps and omissions that caused needless downtime were corrected by documentation and education. Thus, successful implementation of equipment management procedures is the occasion for all staff departments to examine the adequacy of their service procedures to operating departments and start aligning them with the plant manager's conceptualization.

TYPICAL DUTIES OF KEY PERSONNEL

As a general guide to the specific responsibilities and accountabilities of key maintenance personnel, an examination of their typical duties in the support of equipment management is useful.

Maintenance Superintendent or Manager

The maintenance superintendent or manager is responsible for the entire maintenance function. He or she performs this duty through line supervisors.

Superintendents/managers develop an overall maintenance program to ensure that procedures are followed for the identification, requesting, classification, planning, scheduling, assigning, controlling, and measuring of work. This program will be carefully documented to ensure that all plant personnel from manager to workers are educated on its elements. Emphasis will be given to preventive maintenance to ensure that premature failures are avoided and equipment life is optimized. Superintendents/managers will use the most effective and current technology in the maintenance program, with particular attention to condition monitoring. They will organize the maintenance department to ensure effective control of labor and productivity of personnel in their performance of quality work. Superintendents/managers will lead and motivate personnel and develop in them a positive outlook on the effective use of information and its application. They will use maintenance engineers to ensure that the total maintenance program has continuity. In addition, they will ensure that all new construction or installations can be supported by the maintenance program when such work is performed. Superintendents/managers will coordinate carefully with purchasing, warehousing, and shop supervisors to ensure quality material support and effective rebuilding of components. They will utilize maintenance planners to develop plans for resource use on major work. They will diligently coordinate with production managers to ensure that major work is scheduled to ensure the least interruption of operations and best use of maintenance resources. As necessary, they will use administrative personnel to perform tasks related to pay, vacations, absenteeism, and so on.

First-Line Maintenance Supervisor

The first line maintenance supervisor is responsible for everything his or her crew does or fails to do. The supervisor's principal duty is work control and the supervision of his or her crew. The supervisor is responsible for carrying out the assigned work load, including PM services, and unscheduled and emergency work on his or her own initiative. Major planned and scheduled jobs will appear on an approved weekly schedule prepared by a supporting planner and jointly approved by maintenance and operations. These jobs will be converted into daily work plans with the first priority on PM services, followed by the major scheduled jobs and other work fitted into the schedule to coincide with equipment shutdown opportunities. The first-line supervisor is responsible for the conduct of PM services, the control of labor (including overtime), and the procurement of materials for unscheduled and emergency work. Supervisors will assign each crew member a full shift of ready-to-go jobs and control the work effectively. Emphasis will be given to the accurate, complete, and timely reporting of field data by the craftsmen who perform the work. As required, their reporting will be verified. The supervisor will instruct and train crew members, arrange their vacation time, and counsel and discipline them as necessary. Supervisors will prescribe work methods and procedures but will create a work environment to foster initiative and responsibility among crew personnel. Supervisors will ensure the timely completion of work by adequately supervising it.

Maintenance Engineer

The maintenance engineer is responsible to the maintenance superintendent or manager for ensuring the reliability and maintainability of plant equipment and facilities. Specifically, maintenance engineers will devise and implement methods and controls that ensure equipment and facilities are properly installed, correctly modified, properly maintained, and performing effectively. Specific responsibilities include assessment of newly installed equipment to ensure maintainability, parts, maintenance, instructions, prints, adequacy of design, and installation work. In addition, they will monitor ongoing work to ensure that sound craftsmanship procedures are followed. They will also observe the adequacy of the stock of correct parts in proper quantity. Further, they will monitor the quality of parts used to ensure good performance and recommend action to correct inadequacies. Maintenance engineers will review repair history and costs to determine repair, rebuild, overhaul, or corrective maintenance needs. They will also review the conduct of PM services to ensure they are conducted on time, according to standard. Engineers will review equipment inspection deficiencies uncovered as a source of corrective maintenance. They will work with field personnel to develop standards for individual major jobs, establish procedures for performing work, and verify cost information and the quantity of resources used. Periodically, maintenance engineers will make cost and benefit assessments to determine make-or-buy actions. They will also develop recommendations for skills training for craft personnel and training required for planners and supervisors. They will prescribe methods for nondestructive testing and predictive maintenance. Maintenance engineers will also review the information system to ensure its adequacy. They will use information to verify the work load and backlog with the view of making work force level change recommendations.

Maintenance Planner

The maintenance planner provides direct support in planning individual major jobs such as replacing major components or overhauling equipment. In addition, planners may arrange a series of related jobs, the timing of which are part of a scheduled shutdown. Nonmaintenance work such as the installation of new equipment should also be planned if such work is done by maintenance personnel. Planning steps include determining the job scope, planning the job, identifying resources, estimating cost, establishing the logical timing of the work, and coordinating future work timing with operations or engineering (for nonmaintenance work). After work has been scheduled, the planner advises maintenance supervision on the allocation of labor and coordination of field activities such as delivery of materials to work sites. Planners also monitor job execution to determine the nature and timing of ongoing coordination of field support such as the use of cranes, rigging, and transportation. Planners participate in daily coordination meetings during which the timing of work may be shifted to accommodate operating delays, for example. In addition, the planner monitors the conduct of PM services, and reviews repair history and other sources of new work to help determine whether it should be planned.

The maintenance organization should establish criteria for determining which work should be planned. Generally, unscheduled and emergency work should be controlled by field maintenance personnel, except in unusual circumstances. Planners are staff personnel. Thus, they exist to support line supervisors who are ultimately responsible for getting the actual work done. Planners would provide an approved weekly schedule containing competently planned work, scheduled with operations and coordinated to be done at a specific time when there is full agreement that equipment will be shut down. Field supervisors will assign such work to their personnel and add such other unscheduled or running repair work on the same equipment to ensure that every crew or team member has a full shift of bona fide work. In the event that emergency repair work interrupts work on an approved schedule, planners should be advised so that they can readjust the timing of planned and scheduled work to accommodate the interruption. Provision should be made to ensure that labor resources are allocated to planned work, by priority, and to ascertain that the work that has been competently planned and scheduled and is, in fact, accomplished. Procedures should exist so that management and production are advised of schedule compliance. Thereafter, proper corrective actions should be taken. A determination should be made of why work was not completed and the cause identified and reported along with schedule compliance.

Leadman

The leadman is a regular maintenance craftsman who, in consideration of an hourly wage increment and any special talents, can be used temporarily to help a supervisor coordinate elements of major crew jobs. The leadman should not be considered a supervisor since he or she cannot be expected to discipline fellow crew members.

Team Member

A team member is an individual craftsman who shares the responsibility for accomplishing work with other team members.

Team Coordinator

A team coordinator is a rotating coordinator within the team who provides for the control of work during the transition period when a new team is being formed and operating for the first time without a supervisor.

Business Unit Manager

The business unit manager brings together all of the field resources to accomplish day-to-day work under the control of a single person. All of the resources necessary to get the job done are provided to the person responsible and that person is given access to other supporting resources that must be shared by other business unit managers. They would control operating personnel in a specific operation and direct all maintenance personnel required to perform the day-to-day work.

Craftsman (Technician, Tradesman)

A craftsman is an hourly paid maintenance employee possessing specific or combined skills qualifying him or her to perform specific maintenance tasks.

SUMMARY

Key personnel, from executives to plant managers and department managers, must understand the profitability potential of equipment management. Then they must commit to the people and technical improvements necessary to achieve this potential. At each organizational level, the improvement objectives are similar, but the techniques for achieving them are different. Plant managers have the most critical role in establishing the direction of the implementation and aiming their people resources in the right direction. Maintenance and operations managers will witness the greatest exchange of organizational responsibilities, whereas staff managers will take a more active role in equipment selection, logistics, and information. Finally, the craftsman who performs the actual maintenance must be keenly aware of the program that guides his or her actions.

CHAPTER 5

Organization

EFFECTIVE ORGANIZATION CHARACTERISTICS

The organization that carries out equipment management successfully will

- Emphasize cooperative interaction among departments;
- Possess and follow a well-conceived maintenance program;
- Be flexible and responsive;
- Obtain and use information effectively;
- Control resources effectively;
- Have access to additional resources during peak work loads;
- Perform effective planning;
- Provide quality technical support; and
- Communicate effectively with customers, supporters, and managers.

Suitable types of organizations must be weighed against these desirable characteristics. Then the best organization for a particular working environment should be examined and its pros and cons assessed. Next, transitional steps to move from an older, less suitable organization to a more competent one would be established.

CENTRALLY CONTROLLED ORGANIZATIONS

Centrally controlled maintenance organizations lack flexibility and responsiveness. They are best used in subordinate organizations such as lubrication because no one area can use specialized personnel full time. Similarly, a shop solely performing hydraulic component rebuilding would have personnel uniquely qualified operating exclusively in a specially equipped shop.

CRAFT ORGANIZATIONS

The craft organization is the most prominent, enduring industrial maintenance organization. It assigns personnel of the same craft to a single supervisor or puts

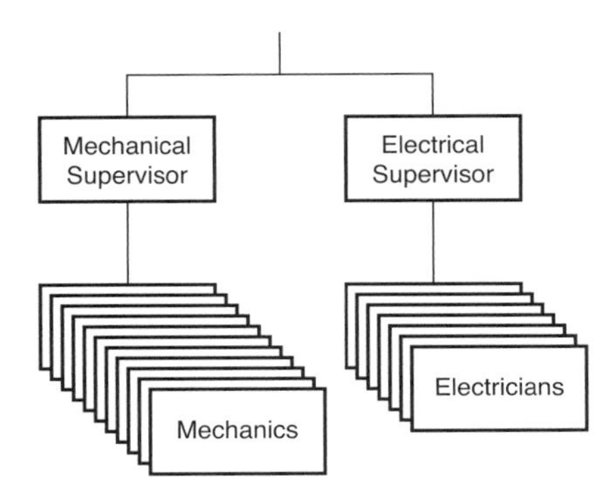

FIGURE 5-1 Personnel distribution in a craft organization

them in the same crew, as in a team (Figure 5-1). Its single-craft orientation means that major jobs requiring additional skills must be drawn from other craft organizations. Thus, the success of many major multicraft jobs performed by craft organizations will depend on the quality of job coordination.

Any time that all personnel are placed within one organization, that organization tends to become the sole decision maker on how the group will be utilized. Then when maintenance needs at the plant level are unknown to the group, the result is often the wrong or least productive use of personnel. Herein is a significant drawback of the craft organization. The most prominent justification for the craft organization is the promotion of the group's craft skill; this is also a weakness of the organization. The boundary lines within which craftsmen apply their skills are invariably too narrow. When compounded by labor contract craft jurisdictions, poor productivity results.

Consider, for example, the task of replacing a simple electric motor. A millwright could easily do the job, but if the labor contract prohibits him from performing electrical work, an electrician must assist. Thus, a simple one-worker ½-hour job costs twice as much, and downtime may be doubled if an electrician is not immediately available. Obviously, the expansion of the jurisdiction of each craft would improve this type of situation. However, this is not always easy to negotiate in a union environment. Moreover, although newly negotiated language might broaden craft jurisdictions, it may not change the way individuals do the work. Thus, old habits and poor productivity remain.

The single-craft alignment of personnel in craft organizations also inhibits the expansion of skills that are required to maintain modern equipment. This is best seen in the traditional separation of mechanical and electrical trades in the craft organization. Although the skills involved are different, there is no reason for the physical separation of crews. Too often, operations wonders why its inoperative equipment must await the visit of the electrician hours after the mechanical group proclaimed its part of the job finished, for example. Any

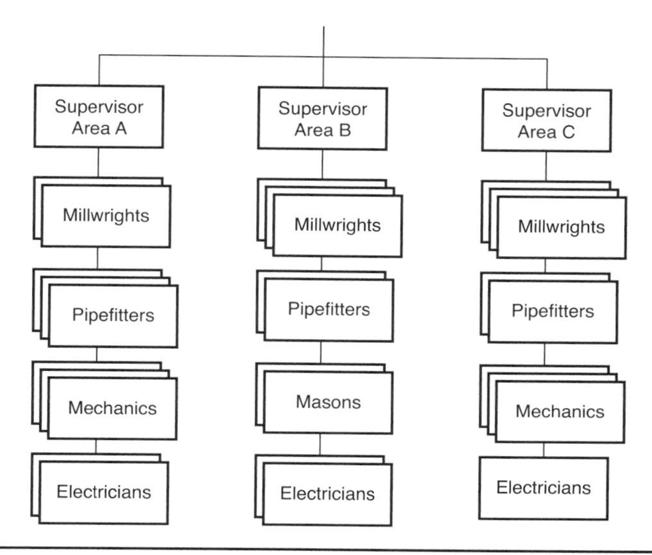

FIGURE 5-2 Personnel distribution in an area organization

organization still committed to this physical separation may not fare well in equipment management. In addition, modern equipment manufacturers will prescribe a total maintenance requirement of several skills for their equipment. If craft personnel perform the prescribed maintenance in the separate, disjointed fashion of craft organizations, they handicap their ability to move into equipment management. Thus, they limit their practical maintenance capabilities by their one-skill orientation. The craft organization is unlikely to be flexible and responsive. More than likely, its capability to apply modern technology may be limited as well. And, while craft organizations may control their own resources properly, they are unlikely to have direct access to all supporting resources.

Instead, there is often considerable "horse trading" among craft supervisors as they haggle over exchanging people to assemble the right skills for a major job. In the meantime, valuable response time is wasted and productivity worsens.

AREA ORGANIZATIONS

An area organization makes one supervisor responsible for all maintenance within a specific geographic plant area. The area organization includes all of the craft personnel necessary to meet day-to-day maintenance needs. Therefore, mechanics, millwrights, and electricians could be in the same crew, depending on the work load of a particular plant area. The area organization successfully eliminates the craft jurisdiction problem of the craft organization (Figure 5-2).

The area organization also suggests that operations would be more satisfied because problems requiring several craft skills can be dealt with sooner. Therefore, for a plant operation, the characteristics of the area organization are more promising than those of the craft organization because of its greater flexibility and responsiveness. Area organizations also have access to other craft personnel required to reinforce them during peak work loads. Direct access to other labor resources required to meet peak work loads suggests that the areas must be supported by other craft groups.

Therefore, these groups must have a variety of craft personnel to meet the diverse needs of peak work loads of several areas concurrently. Consequently, the allocation of labor resources is more likely to meet plantwide priorities than is the craft organization.

Typically, plants wishing to be able to reinforce field areas with additional craft personnel during peak work-load periods prefer the area organization. They find that the clear definition of responsibilities is not only more effective but it is more compatible with the objectives of equipment management as well. The most common support organization is a craft pool. Personnel from the craft pool are awarded to field areas experiencing peak work loads, like a periodic shutdown, based on plant-level work priorities.

In contrasting the craft and area organizations, the best choice in terms of flexibility and responsiveness would be the area organization. However, the quality of planning and the use of technology and quality information are also important considerations.

BUSINESS UNITS

A business unit, in which one person is responsible for both operations and maintenance, is a very practical and workable organization for carrying out equipment management. Aside from removing the obvious "finger pointing" between operations and maintenance, the business unit builds leadership and professionalism. If, for example, the unit manager is a long-term production type, she can aid the development of greater professionalism among craft personnel. They depend on her to lead and she is fully dependent on them for quality work. Thus, the business unit manager who exhibits quality leadership will realize professional support from craft personnel in return. In the example in Figure 5-3, the business unit manager controls both operations and maintenance. If the maintenance activity is large and complex, she may utilize a planner exclusively. Otherwise, planning services would be made available as needed.

The business unit also brings equipment operators and craft personnel into closer working proximity. Thus, there is opportunity for them to create a maintenance–operations team as well. Soon, the capabilities of operators to perform specific, helpful maintenance tasks become apparent. Similarly,

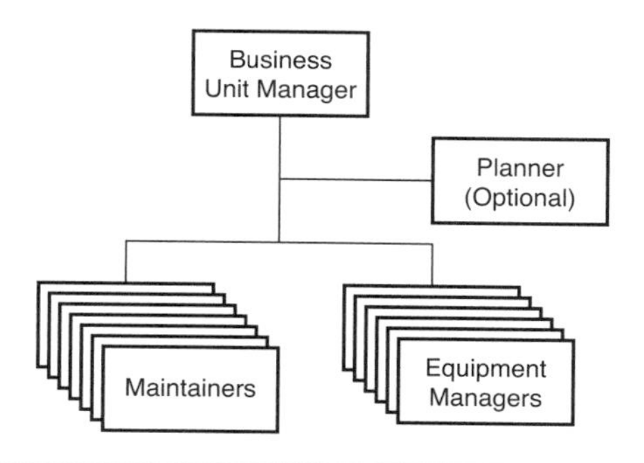

FIGURE 5-3 Business unit organization

equipment operators learn more about the way their equipment works and are better able to identify potential problems. As the maintenance–operations team effort grows, the equipment operators see maintenance personnel in a different light. They are more confident, and they report problems promptly and correctly. Maintenance personnel, in turn, volunteer shortcuts in adjusting or calibrating equipment. Soon, a real team environment emerges. Business units are a potentially solid way to support equipment management.

PLANNING AND SCHEDULING

Maintenance planners are often utilized in larger organizations where there is considerable major work requiring detailed, advance preparation before the work can be scheduled. Smaller organizations still have a need for planning, but the task is often carried out by supervisors or team members. For those organizations using planners, it is important to establish what the planner should do, how it should be done, and what tools are required.

The principal task of the maintenance planner is to organize selected major jobs in advance. Then when the work is carried out, it can be performed in the most efficient way. It will use the least amount of labor and be completed in less elapsed downtime. In addition, the more deliberate way in which work is done will ensure quality work.

Although it can be argued that all work requires some degree of planning, there is little value in sending all work to a planner. What value is there in sending known emergency repairs to a planner? Planners cannot help. Nor does it make sense to send a multitude of small jobs to a planner when, clearly, no planning is necessary. Organizations that funnel all work through planners simply convert the planner to a clerk status and lose planning help in the pre-organization of major jobs and the obvious benefits. There should be practical criteria for determining which work requires planning. Criteria should be developed jointly by planners and field personnel to ensure fewer disagreements. Criteria

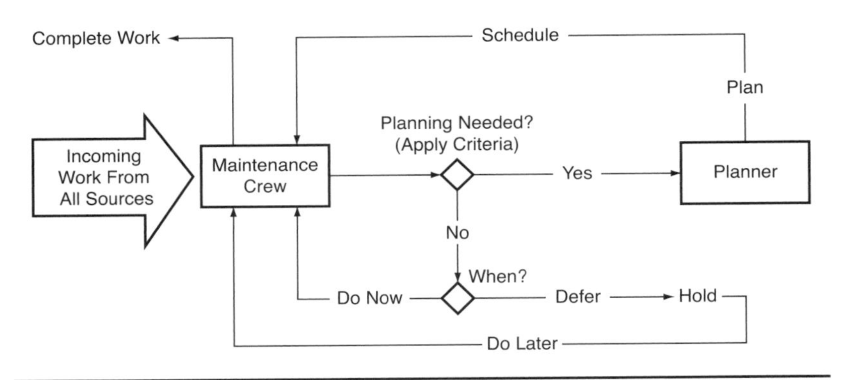

FIGURE 5-4 Clear division of work and control between maintenance crew and planner

also help to establish a clear distinction between the work that field crews would handle and work sent to the planner. Experience guides the field crew in the use of the planner. Agreement on a criteria as to which work requires planning quickly converts potential controversy into a businesslike activity. For example, if the crew is to control, schedule, and do certain work, then they must have direct access and availability to materials needed, drawings, schematics, and tools. They must also coordinate equipment shutdown. If, on the other hand, materials must come from several sources, shop work is required, or a lengthy shutdown period must be negotiated with operations, then such work should be sent to the planner (Figure 5-4).

Scheduling of routine, repetitive preventive maintenance (PM) services requires no planning. The whole PM program was planned when it was set up. This included specifying services, developing the checklists, and establishing the repetitive frequency of services. Thereafter, PM services are only scheduled repetitively.

Typically, PM services for multiple units of mobile equipment can be easily scheduled by entering the latest operating hours into the computer once a week. The computer does the rest and even prints the checklist on request. Thus, a field organization can easily schedule dynamic PM services without the involvement of a planner.

Depending on the size of the total maintenance organization, the number of planners can range from a few individuals to an entire planning staff. Generally, the number of planners is related to the volume of planned work and the number and variety of personnel carrying out the planned work. Ideally, 60% of the total maintenance work force should be performing planned/scheduled maintenance. By multiplying the correctly established work force size by 0.60 and dividing that number by 15 (the planner–craftsman ratio), the number of planners needed can be approximated. For example,

> In a work force with a 150 maintenance workers, 60% would perform planned/scheduled maintenance. Therefore, the number of personnel doing this type if work would be 150 × 0.60, or 90 workers. Dividing 90 by 15 would yield a requirement for 6 planners.

Planned and scheduled maintenance is invariably done with 12%–15% less labor and at least 6% less downtime when compared with other similar unplanned jobs.

Jobs that require replacing of major components are performed on a regular, periodic basis. Therefore, development and use of standard task lists, bills of materials, and tool lists can simplify planning when compared with unique jobs planned from scratch. With periodic maintenance and the use of standards, most jobs are 90% planned and require only slight adjustments, as similar jobs are repeated on different units of the same type of equipment.

A significant amount of the planner's time is spent obtaining materials. On average, a planner can spend 40% of his time identifying and procuring materials. Many large maintenance organizations have improved planning performance considerably by assigning material coordinators to the planning staff. The material coordinators assume responsibility for materials procurement after the materials are identified. Thus, planners gain time for other planning tasks. Material coordinators should be material control specialists. Most are placed on special assignment from the warehouse or purchasing department. This reinforces the desirable interdepartmental cooperation required in equipment management.

TECHNICAL SUPPORT

Successful equipment management will require effective application of condition-monitoring techniques. These are necessary to ensure that equipment condition is known at all times and that the organization can respond rapidly and correctly to preclude the consequences of failure. The skill with which the right techniques are identified and applied and their data analyzed will influence the success of equipment management. In major industrial plants, the variety and quantity of equipment that must be monitored will require a technical support organization that can effectively administer a multitude of technologies. These larger organizations have established maintenance engineering organizations that ensure the reliability and maintainability of equipment through the application of modern technology. The use of a combined maintenance engineering staff (Figure 5-5) is a potentially effective way of coordinating planning and technical support for equipment management. It also reduces duplicate staff and focuses on accountability as with business unit managers.

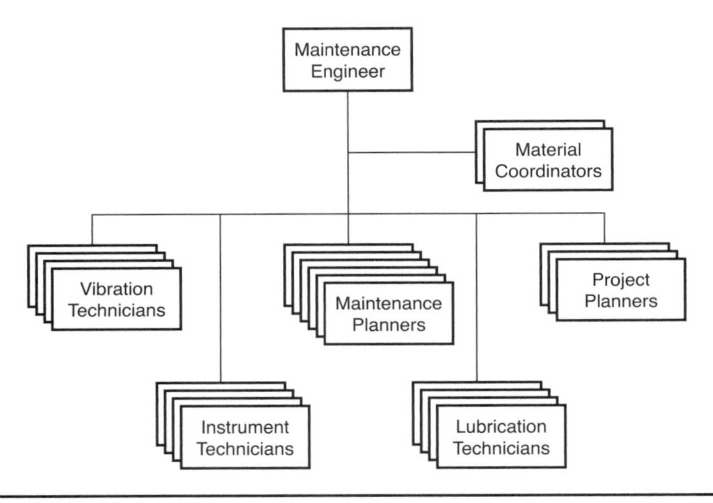

FIGURE 5-5 Combined maintenance engineering staff

ORGANIZATIONAL CHANGE

Organizational change often invokes resistance. Affected personnel must be convinced that the changes are beneficial, first to them as individuals, then to their immediate work groups (crews), and finally to the plant. They are always aware of pending changes, and most are personally concerned with the outcome. However, if they are not informed about the nature of the changes, they will be uncertain and confused. Thus, they could resist a potentially beneficial change that they would otherwise support if they knew more about it. The most effective ways to overcome resistance to change are

- Education—Let people know what the changes are and how they will be affected.
- Participation—Give personnel an opportunity to think through the changes and make alternate proposals that will get the job done.
- Time—Give people an opportunity to absorb the changes.
- Security—Change will not be successful unless both salaried and hourly personnel are assured that their jobs, thus their economic security, will remain intact (Judson 1966).

Resistance to change is always in proportion to the degree of knowledge about the changes. Therefore, the sooner the education process begins, the greater are the chances of overcoming resistance and getting on with beneficial changes. Thus, it is important for the plant to establish how the changes will be implemented. Getting people involved and securing their informed participation and support will always yield more positive support for any organizational change.

IMPLEMENTING TEAM ORGANIZATIONS

The visualization of smaller, more productive work forces using teams, operating without supervision and working efficiently to yield high-quality work at lower cost, has great appeal to industrial organizations. They visualize better work control through employee empowerment and reduced operating cost with fewer people and potentially greater productivity. However, moving successfully from a traditional industrial maintenance organization such as a craft organization to a team organization is a process that must be carefully prepared for and expertly implemented. This section outlines the steps necessary to form a successful team organization.

Step 1. Team members must be convinced that the team organization will make their work easier and more effective. The team organization has a better chance of success if individuals are convinced that the team will help them to operate more effectively, produce quality work, and bring greater satisfaction. They will always ask, "What's in it for me?" Managers must be able to answer this question convincingly. It will not be enough to point out the successes of other companies. Better quality work, by itself, is rarely an exciting enough objective to cause the average American worker to embrace a team organization. But as dramatic changes in plant operation are observed when employees are forced to either buy and operate an unprofitable plant or become unemployed, one gets closer to the type of motivation that can be emulated. Similarly, maintenance employees on back shifts, without supervision, have been seen hustling to get work done. Thus, the productivity potential is there. One must now find out how to mobilize it.

Step 2. Ensure direct employee participation from the very first concept meeting to the final implementation and beyond. Employee productivity, improved performance, and cost reduction are among the expectations of management for a team organization. However, management also knows that the cost of maintenance will not go down unless the number of people doing it and the frequency at which it is done are reduced. Employees are also aware of these facts. In a union environment, immediate opposition to fewer people will be a strong, adverse reaction to the idea of teams. But instead of excluding the union from management discussions, they can be made part of the discussion. A credible explanation can be offered as to why a team is thought to be necessary.

If the plant work force must get smaller to reduce costs, then some jobs can be preserved by displacing outside contractors with current employees. If the work force simply must be reduced, the difficult financial circumstances the company must overcome can be explained. If early retirement is being considered, the options available can be divulged. It is much better to have an open discussion with groups that have opposing views and know their objections than to have groups sabotage the effort with hidden, misinformed actions. When the open discussion is maintained, most objections can be overcome through education.

Step 3. Make certain that the maintenance program is sensible, clearly documented, and well-understood by all plant personnel. Successful implementation of a team organization requires a well-defined maintenance program. With the framework of a solid program, craft personnel and operators will more readily identify and respond to new work and perform ongoing tasks more effectively. Craftsmen, for example, have significant experience in diagnosis and repair but little experience in work control. Their supervisors, now gone, did all of this for them. This void must be filled before the team concept can be implemented. If the new team does not have a clear program to follow, especially during the initial stages of team implementation, problems can mount quickly. Typically, when bombarded with requests for emergency repairs, the new team with a poorly defined program responds to the most threatening tasks and then may retreat to the sanctuary of their lunch room to let the dust settle. While many team members appreciate that condition monitoring can prevent emergencies, it is simply not their responsibility to create the monitoring program. Such a program should have existed prior to team formation. Thus, organizations aspiring to successfully implement a team must bring the team to life within the framework of a well-defined equipment management program.

Step 4. Because the supervisory position from the traditional organization will become redundant, phase out this position by creating a more important task for the supervisor's considerable talents and experience. Often, the biggest surprise in team implementation is the potential resistance from the former maintenance supervisor who may see himself becoming redundant. With an uncertain future, he may fight to preserve the status quo and delay the formation of the team. When converting from a traditional organization to a team organization, the supervisor must be assured of his future task in the team organization.

Good opportunities for the ex-supervisor exist in planning, coordination, and, depending on technical background, the conduct of maintenance engineering tasks such as condition monitoring, equipment modification, or the development of standards.

The transition to a team organization may be the first time that crew members are asked to operate without a supervisor. Often a gradual transition reduces much of the trial and error as new team members cope with the logistics of work control required by the maintenance program. The use of a rotating team coordinator helps to bridge the transition. As the new team is launched, one team member is designated as the team coordinator. For the next several weeks, the team coordinator makes the necessary logistics-related decisions, such as arranging for the on-site delivery of materials for a major job. During this initial period, the team coordinator explains the procedures that he would recommend in similar situations. Other team members question and discuss the procedures and, if necessary, add them to the documented maintenance program. Team members rotate through the team coordinator's position until all team members have had an opportunity to act as a coordinator. By the time that all team members have participated, most logistical and administrative procedures will have become standardized and accepted.

Step 5. If a team is being assembled in a new plant, establish the desired qualifications for team members and recruit new team members carefully. Generally, organizations that recruit directly into a new nonunion organization stand the best chance of success. Their personnel directors are able to spot and reject the "storm trooper" type in favor of the potential team player. As a result, the team starts out with fewer handicaps. Despite this, many "frustrated contractors" are hired as craftsmen.

For example, a millwright with considerable experience in performing construction work is not, necessarily, a maintenance millwright. These types of personnel do not look forward to diagnosing a serious equipment problem while lying in the mud. Instead, they would much rather be involved in a "project" with new materials, plans, drawings, and a comfortable, no-pressure timetable in which to complete the work.

When "contractor-type" craft personnel are substituted for "maintenance-type" craft personnel, vital PM services that help guarantee reliable equipment may be displaced with questionable projects encouraged or even solicited by these individuals. Recruiters should also look for personnel with good craft skills who exhibit potential for team membership. If prospective team members appreciate the framework of maintenance management, their contribution to a successful team is much greater. By starting with the right type of people, the plant can avoid taking on difficult behavioral changes that are required to make the new team effective.

Step 6. Carefully weigh whether the new team will be paid by salary or an hourly wage. The maintenance team may be made up of former hourly workers whose status may now be changed to that of salaried employees. These new salaried employees may be looking for the perks (e.g., flexible lunch periods) that other salaried employees seem to enjoy.

Employees must realize that their work status and shift assignments may not permit them the same latitude in time on or off the job afforded other salaried personnel. Equipment maintenance needs do not recognize individual preferences. When the need for maintenance work arises, there must be a rapid, dependable response. To help team personnel convert from hourly wage to salaried status, ground rules must be provided to clarify expectations.

Step 7. Establish team decision-making parameters carefully. A team may be qualified to make decisions about how to perform work. But when they are asked to decide how to manage maintenance or which shifts they will work, they typically cannot deliver the answers. Therefore, the boundaries within which the teams can make decisions should be determined in advance. The team can schedule established PM services after coordinating with production, but they cannot specify what services are to be performed. They could, however, recommend changes to existing programs. They can modify equipment, but only as prescribed by maintenance engineering, and they can recommend how the modifications might be made.

Step 8. Determine how craft skills will be evaluated and remedial training provided to improve lagging skills. A team could become a hiding place for the worker who needs skills training. Similarly, others might become whistle-blowers, anxious to point out someone who doesn't carry his or her share of the work. These situations can demoralize the team's overall efforts and should be precluded before they can develop. Therefore, team development and implementation should include methods for determining training needs and followup corrective training. Continuous skills training should be provided, but procedures should also be established for evaluation of skills and provision of remedial training when necessary.

> One newly formed team established a two-level program to identify and correct training needs. By linking skills with incremental wage increases, craft personnel were able to request consideration for an increase. After a request was made, a group of the candidate's peers assembled an informal group to judge the candidate's knowledge for the new level sought. Questioning was tough and fair. If the candidate met the interview requirement, he or she was then tested for practical experience. A second group had rigged a unit of equipment in the plant with several specific faults. The candidate was given an opportunity to diagnose and correct all of the faults and restore the equipment to satisfactory condition within a specific period of time. If both the interview and the practical experience requirements were met, the candidate was endorsed for the new status.

Step 9. Give the team firm work control procedures, as most team members have not had the training or the experience to develop their own. Within a framework of reasonable work control procedures, such as making daily work assignments or determining which type of work requires planning, most teams can be successful. To illustrate,

> At a multiple-shift manufacturing plant, maintenance personnel on the second and third shifts operated without maintenance supervision. Instead, operations supervisors provided general guidance as required but expected the maintenance people to get the work done. This arrangement thrust craft team members into a situation requiring that they "call their own shots." Fortunately, they had received considerable training on the overall maintenance program and, as a result, saw themselves as the direct maintenance contacts on the back shifts. With this education, they saw the total picture of work control

> and were able to competently manage work requests and unscheduled repairs, and handle emergencies. They did this with praise and compliments from their operations counterparts.

Step 10. Organize maintenance and establish a working relationship with operations that fosters team utilization and development. A concept of how operations and maintenance personnel would work together should be included in team training with emphasis on functioning as an interdepartmental team.

> In an underground coal mining operation, each six-person production crew was assigned one electrician. The electricians realized their success would depend on working as a team with the operators on the crew. They not only made it clear to operators what services they would perform but when they must be done. To solidify the team working relationship, electricians tried to be helpful to operators. If an operator needed a break, for example, an electrician relieved him or her. When the operator returned to the unit, the electrician provided pointers on checking equipment while it was being operated. In turn, if there were heavy repair jobs to be done, the electrician could count on the operators for help. No one told them to do it; they just did it. This type of maintenance–operations cooperation made the team successful.

Team members will have had little direct experience in daily plant operation, yet they will inherit a responsibility to support it. A well-defined maintenance program will avoid much trial and error. Sensible guidelines on activities such as decision making and interdepartmental interaction will also help. Continuous skill training should continue so that teams will continue to produce quality work and gain respect. Supervisors must be given a clear picture of future roles so they can effectively support the team members they used to control. The environment in which the team is launched should offer a spirit of interdepartmental cooperation, reinforced with supportive management actions during transition into the team organization.

SUMMARY

Successful implementation of equipment management is built on the concept of cooperative interaction of departments and individuals. Therefore, any organizational change made to accommodate the implementation of equipment management must reinforce this concept. The new organization must be flexible and responsive. It must be able to apply modern technology efficiently, then obtain and use information effectively. It should also have the capability to control maintenance resources at the lowest organizational level and have direct access to other resources needed to support field operations during peak work loads. In addition, the organization must get under way with a well-conceived maintenance program. This program must emphasize solid planning and effective technical support for field operations. Finally, the new organization must be able to communicate well with customers, supporters, and managers. In doing all of these things, it can more effectively support equipment management.

REFERENCE

Judson, A.S. 1966. *A Manager's Guide to Making Changes*. New York: John Wiley & Sons. pp. 68–85.

CHAPTER 6

Work Load Versus Work Force

WHAT IS THE WORK LOAD?

The work load is a measure of the amount of essential work performed by maintenance. Definition of the individual categories of work (preventive maintenance versus emergency repairs, for example) that make up the work load helps maintenance to determine the proper size and craft composition of the work force. The categories of work should be carefully defined to ensure clarity and common understanding. This avoids the potential confusion should operations and maintenance, for instance, have opposing ideas of what constitutes an emergency repair.

CONTROL OF WORK

Each category of work is controlled differently. For example, preventive maintenance (PM) services are scheduled week by week and done routinely (same checklist) and repetitively (every 4 weeks or every 750 operating hours). Emergency repairs require an immediate response. Planned maintenance is deliberately scheduled after joint operations–maintenance agreement as to when it will be done.

CONVERTING WORK LOAD TO WORK FORCE

Establishing the right number of personnel required is an essential purpose of work load determination. For instance, planned work is estimated by craft. When major components are replaced at known intervals, labor requirements can be determined. The end result is a reasonably accurate estimate of the labor required for this type of planned work. Labor for routine maintenance like shop cleanup or training can be allocated. But emergency and unscheduled repairs can only be limited by effective preventive maintenance. In addition, backlog data (estimated man-hours by craft) can be used effectively to adjust the size and composition of the work force as the work load changes.

Prudent plant managers know that the cost of maintenance will not go down unless it is done with fewer people and less often. They can take the first step toward ensuring that the right work force size is established by insisting that the work load be clearly identified. Over the years, maintenance has been casual in determining how many personnel they need. Often, they guessed at the correct number. For instance,

> One senior maintenance manager casually announced that his technique for determining how many personnel were needed amounted to, "I keep adding people until the overtime goes down."
>
> Another suggested that in his mobile equipment shop he needed "0.6 men for each engine." But he had no idea where this mysterious relationship came from.

Maintenance has, in the past, not been eager to commit to a work force level nor have they been keen to measure productivity to assess the quality of labor control. There is considerable evidence that effective labor control, the objective of good productivity, has never been a solid maintenance trait. Why else would they assign all of their personnel to the first-line supervisor whose basic view is that "14 are better than 9 when we get surprises"? Why also has maintenance stuck with craft organizations in which one supervisor controls all personnel of a single craft—knowing that this is the least productive use of labor? Similarly, there is little rationale for the traditional separation of electrical and mechanical skills if the organization is serious about effective labor control. This mechanical–electrical separation often causes operating supervisors to feel like they are dealing with two separate maintenance departments. The resulting delays may further convince some operating personnel that their maintenance organization is disinterested in effective control of labor.

WORK LOAD CATEGORIES

Maintenance work should not be oversimplified by limiting it to preventive and corrective maintenance. Corrective maintenance, for example, is made up of emergency repairs, unscheduled repairs, and planned/scheduled maintenance. The work load definition begins with a definition of *maintenance*:

> Maintenance is the repair and upkeep of existing equipment, facilities, buildings, or areas in accordance with current design specifications to keep them in a safe, effective condition while meeting their intended purposes.

Maintenance and nonmaintenance work should not be mixed together. Equipment modification, installation, and construction work are not maintenance and are often capitalized. Nonmaintenance work should be carefully identified and labor allocated to it only after basic maintenance needs are met. Figure 6–1 depicts the overall maintenance work load made up of corrective

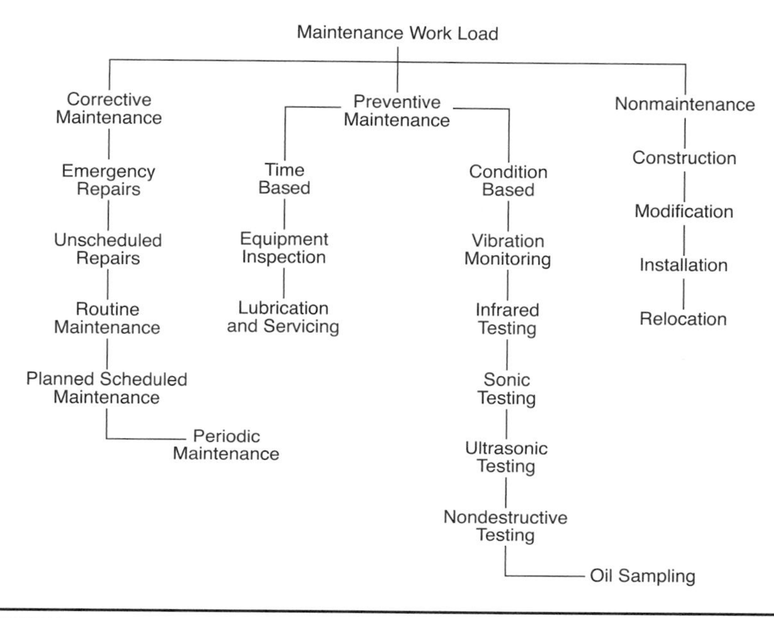

FIGURE 6-1 Maintenance work load

maintenance, preventive maintenance, and nonmaintenance work, and the subdivisions of each.

Preventive Maintenance

Equipment inspection, testing, and condition monitoring are performed to ensure the early detection of equipment deficiencies. Lubrication, cleaning, adjusting, calibration, and minor component replacements (belts, filters, etc.) are performed to extend equipment life. An effective PM program will feature the detection-orientation of inspections, testing, and condition monitoring to avoid premature equipment failures and create the lead time to plan the resulting work. These actions will then displace the need to react to emergency repairs created by equipment failures. The PM work load can be computed. For example, a PM inspection done every 2 weeks requiring 3 mechanical man-hours per repetition would require 78 mechanic man-hours annually. Similarly, estimated craft man-hours for all PM services would yield a total craft labor estimate for the entire PM work load. Providing PM services are defined as suggested above, and they include only the services specified, not the new work that inspections, monitoring, and testing generate. A plant should spend between 9% and 11% of its total work load on PM services. The PM schedule compliance should be targeted at 85%.

Corrective Maintenance

Emergency Repairs. Emergency repairs are repairs needed as a result of failure or stoppage of critical equipment during a scheduled operating period. Danger to personnel, extensive further damage to equipment, or substantial production loss will result if repairs are not made quickly. Emergency repair work loads cannot be computed in advance. But they can be limited by a good PM program that uncovers equipment deficiencies before equipment fails. This allows more work to be planned. Generally, when PM services are carried out faithfully, emergency repairs should not exceed 8%–10% of the total work load.

Unscheduled Repairs. Unscheduled repairs are defined as nonemergency work of short duration that can be accomplished any time within about a week with little danger of equipment deterioration in the interim. Most unscheduled work is performed by one person in 2 hours or less. Generally, this type of work uses common shop stock or easily available stocked or free-issue items. All such minor work should be controlled by the maintenance crew that has discovered the work without the need to obtain the assistance of a planner. The work load of unscheduled repair work cannot be computed in advance. Generally, unscheduled repairs should not exceed 12%–15% of the total work load.

Routine Maintenance. Routine maintenance involves activities such as janitorial work, building and grounds work, training, safety meetings or shop cleanup, and highly repetitive work such as tool sharpening. Routine maintenance activities can account for 5%–6% of the total man-hours available for maintenance. Therefore, this portion of the work load must be specified and allowances established. Activities making up the routine maintenance work load can be defined and specified. For example,

- Each electrical craftsman will undergo 65 hours of training annually.
- One full-time utility person is required for tool repair.
- Each shop craftsman is allowed 15 minutes per day for cleanup.

The routine maintenance portion of the work load can be measured, regulated, and controlled through the use of a standing work order against which labor is reported. The actual use of labor for activities like shop cleanup or training can then be compared with standards to bring this use closer to desired levels.

Planned/Scheduled Maintenance. Planned/scheduled maintenance is extensive major maintenance whose scope, cost, complexity, and importance warrant advance preparation and timely scheduling of work. Properly planned and scheduled maintenance can be carried out efficiently in the least amount of elapsed downtime, with the most effective use of resources and the highest assurance of quality work. Typically, it includes major jobs such as the replacement of major components (engines, drive motors, transmissions, power trains, etc.) or the overhaul of a total unit of equipment by replacing all major components and restoring it to a new, as-designed condition. This work also includes unique major jobs. All such work requires advanced planning, lead time to assemble materials, and the scheduling of equipment shutdown to ensure availability of maintenance resources including labor, materials, tools, and shop facilities.

Scheduling should include advance agreement with production and operating personnel that the work will be carried out at a time that least interferes with their operations while enabling maintenance to schedule their resources most advantageously.

Nonmaintenance (Project Work)

Project work is made up of construction, installation, relocation, or modification of equipment, buildings, facilities, and utilities. The nonmaintenance work load is determined in the same way as is the work load for planned/scheduled maintenance. That is, because most of this type of work will be planned, estimates can be made to determine the labor required by craft. Then, historically, projections can be made on the labor required for this work load by observing backlog trends kept separately for nonmaintenance work. Typically, nonmaintenance elements are defined as follows:

- Construction—The creation of a new facility or the changing of the configuration or capacity of a building, facility, or utility.
- Modification—The major changing of an existing unit of equipment or a facility from original design specifications.
- Installation—The installation of new equipment.
- Relocation—Repositioning of major equipment to perform the same function in a new location.

ESTABLISHING WORK FORCE SIZE AND COMPOSITION

Consider these eight steps in establishing the proper size and craft composition of the work force:

1. Establish a policy requiring that the work load be measured.
2. Carefully identify and define the work required of maintenance and acknowledge the implications of how the work is controlled and performed.
3. Measure or estimate the labor required to perform each category of work.
4. Assemble the preliminary work load data.
5. Establish a target distribution of labor and compare actual labor use with it.

6. Determine the number of supervisors required.
7. Assess the quality of labor control.
8. Watch and correct labor control performance trends.

Step 1

Establish a policy requiring that the work load be measured. Do not expect maintenance to do it voluntarily. They will not rob themselves of flexibility to respond to labor demands they can't control. Therefore, consider these policies:

- The maintenance work load will be measured on a regular basis to help determine the proper size and craft composition of the work force.
- The productivity of maintenance will be measured on a regular, continuing basis to monitor progress in improving the control of labor.
- Every effort will be made to implement and utilize organizational and management techniques to ensure the most productive use of maintenance personnel.

Step 2

Carefully identify and define the work required of maintenance and acknowledge the implications of how the work is controlled and performed.

Step 3

Measure the labor required to perform each category of work. PM services, planned/scheduled maintenance, routine maintenance, and nonmaintenance (project work) work loads can be calculated or estimated. Unscheduled and emergency repairs are unexpected and they cannot be effectively measured or estimated in advance. Rather, they are able to be limited by better PM efforts to detect equipment problems and respond with more planned work. Any planned/scheduled maintenance work that is repeated at regular intervals (like major component replacements) provides an additional way to determine the labor required. Because these events have been performed before, each repetition yields more data on actual labor use by craft. By associating each event with the projected time it may be repeated and the number of man-hours by craft required, the work load for the entire component replacement program can be estimated. In many heavy industries, the major component replacement program represents between 60% and 65% of all the work that can be planned and scheduled. Unique major jobs requiring planning have no historical basis except that they may be peculiar to a specific process, environment, or type of operation. Thus, the portion of the planned/scheduled maintenance work load representing unique major jobs can be factored in by experience and the examination of historical records.

TABLE 6-1 Man-hours required per craft

Cost Center 6	Crafts					
Type of Work	Millwright	Mechanic	Electrician	Welder	Laborer	Total
Preventive maintenance	—	1,260	1,080	—	—	2,340
Scheduled	3,060	2,160	900	2,880	2,700	11,700
Unscheduled	1,620	1,620	360	540	540	4,680
Routine	—	1,260	1,080	—	—	2,340
Emergency	720	900	180	180	360	2,340
Total man-hours	5,400	7,200	3,600	3,600	3,600	23,400

TABLE 6-2 Workers required per craft

Cost Center 6	Crafts					
Type of Work	Millwright	Mechanic	Electrician	Welder	Laborer	Total
Preventive maintenance	—	0.7	0.6	—	—	1.3
Scheduled	1.7	1.2	0.5	1.6	1.5	6.5
Unscheduled	0.9	0.9	0.2	0.3	0.3	2.6
Routine	—	0.7	0.6	—	—	1.3
Emergency	0.4	0.5	0.1	0.1	0.2	1.3
Total man-hours	3.0	4.0	2.0	2.0	2.0	13.0

Step 4

Assemble the preliminary work load data into a logical format. Table 6-1 illustrates a suggested format by which estimated man-hours by craft can be tabulated against the category of work performed. By assembling estimated craft man-hours by type of work, an overview of needs can be determined. Next, convert the man-hour estimates from Table 6-1 into an equivalent number of craft personnel by dividing each by 1,800 man-hours per person per year (Table 6-2).

The result is that a preliminary work force of 13 workers made up of 3 millwrights, 4 mechanics, 2 electricians, 2 welders, 2 laborers (and a supervisor if a traditional organization is used) would be required to perform maintenance in Cost Center 6 with a craft distribution as shown in Table 6-2.

Step 5

Establish a target distribution of labor and compare it with actual labor use. Take corrective actions, like more PM emphasis, to align actual levels with target levels. Based on the previous examples, having established a preliminary crew of 13 made up of five different crafts, labor utilization data is used to confirm craft man-hours worked against categories of work. A typical, desirable distribution of labor by category of work is shown in Table 6-3.

Using labor utilization information, the actual distribution of craft man-hours is compared with target levels (Table 6-3). Data such as shown in Table 6-4 are obtained by reporting labor against the categories of work on which the data were used. Against the established targets, the actual use of man-hours can be brought closer with more emphasis on PM services and a reduction of the nonmaintenance work being performed. More preventive maintenance will reduce the man-hours spent on unscheduled and emergency repairs and allow this labor to be available for planned/scheduled maintenance. As more data are gathered and performance comes closer to the target distribution, the work load can be confirmed. See Table 6-5.

TABLE 6-3 Target distribution of labor by work category

Type of Work	Target, %
Preventive maintenance	11
Planned/scheduled maintenance	46
Nonmaintenance (part of planned/scheduled maintenance)	7
Unscheduled repairs	19
Routine activities	9
Emergency repairs	8

TABLE 6-4 Reported work load

Type of Work	Target, %	Actual, %	Difference
Preventive maintenance	11	3	–8
Planned/scheduled maintenance	46	27	–19
Nonmaintenance	7	20	+13
Unscheduled repairs	19	21	+2
Routine activities	9	9	0
Emergency repairs	8	20	+12

TABLE 6-5 Revised work load

Type of Work	Target, %	Actual, %	Difference
Preventive maintenance	11	9	–2
Planned/scheduled maintenance	46	41	–5
Nonmaintenance	7	10	+3
Unscheduled repairs	19	21	+2
Routine activities	9	9	0
Emergency repairs	8	10	+2

Step 6

Determine the number of supervisors required. If a traditional maintenance organization with a supervisor and crew is used, the number of supervisors required must be determined so that control of the crew members' activities can be ensured. To establish the number of supervisors required, first determine the shifts requiring supervisory coverage:

Areas	Annual Shift Coverage	Shifts per Year
3	52 weeks × 5 shifts per week	780
4	52 weeks × 7 shifts per week	1,456
Total shift coverage required		2,236

In this example, three areas require coverage for five shifts per week and four areas require coverage for seven shifts per week. The total annual shift coverage is 2,236.

Next, determine the shift coverage capability per supervisor:

5 shifts per week × (52 weeks – 7 weeks of vacation, holiday, and compensatory time) = 225 shifts coverage per supervisor per year

Now the total shift coverage required is divided by the shift coverage capability of a single supervisor:

2,236 shifts/225 shifts per year per supervisor = 9.9, or 10 supervisors

Because there are only seven areas, there should be seven full-time supervisors and three relief supervisors. Using this data as a guide, the maintenance organization must then decide whether some shifts actually require a maintenance supervisor. For example, a plant with a small crew on each afternoon and night shift might place the crews under the operational control of the operations shift supervisor. However, if this is done, the operations supervisor must be aware of the work the crew has been assigned and under what circumstances he or she will interrupt their assigned tasks (e.g., emergencies only). Similarly, crew members must understand this working relationship and cooperate.

Step 7

Assess the quality of labor control. When adequate man-hours are spent on preventive maintenance, the results are

- Less labor used on emergency and unscheduled repairs,
- More labor used on planned/scheduled maintenance,
- Better organized planned work, and
- Work performed more productively.

TABLE 6-6 Maintenance labor utilization report

Week 40—Ending September 30
Department 207

Craft	Preventive Maintenance	Scheduled Maintenance	Unscheduled Repairs	Emergency Repairs	Non-maintenance	Total Work Force
Millwright	55	276	102	88	67	588
Mechanic	46	244	96	66	42	494
Electrician	23	122	54	32	46	277
Welder		72	58	21	66	217
Instrument	18	32	30	12	45	137
Pipefitter		48	12	9	12	81
Laborer		121	6	5	41	173
Total man-hours	142	915	358	233	319	1,967
% Distribution	7	47	18	12	16	100

Therefore, weekly trends in the use of labor by category of work are an excellent means of observing the impact of more preventive maintenance on reducing emergency repairs and improving the amount and quality of planned/scheduled maintenance. However, the traditional problem of getting craft personnel to accurately report the use of actual man-hours will require exceptional leadership, even in an era when most information systems can handle this data competently. In the illustrative report of maintenance labor utilization shown in Table 6-6, when 7% of the maintenance labor was spent on preventive maintenance, then 47% was spent on planned/scheduled maintenance with 18% on unscheduled repairs, 12% on emergency repairs, and 16% on nonmaintenance projects.

Step 8

Watch labor control performance trends. From the plant manager level, an excellent source of information on labor control efficiency is the cost of labor to install each dollar of material. Costs well above the trend will identify crews that are controlling labor poorly. Such crews should be subjected to closer scrutiny in the way they operate. Most likely, there is a poor PM effort, little planning, and excessive emergency repairs.

USING BACKLOG INFORMATION

The backlog provides a means by which maintenance can adjust the size and composition of the work force as the work load changes. The backlog is the number of estimated man-hours of all identified but incomplete planned and scheduled work. It identifies the degree to which maintenance is keeping up with the generation of new work and permits maintenance to adjust the size and composition of the work force as the work load changes. The backlog is explained in more detail in Chapter 13, Essential Information.

SUMMARY

The era of intense competition among industries to capture market share has seen many companies moving their manufacturing entities overseas to take advantage of lower labor costs. Yet the domestic industrial base cannot surrender command of vital, strategic industries simply because of a higher living standard. Instead, they must seek ways to become more competitive by improving labor productivity. The first step in this process is the accurate identification of the work load and, from it, the accurate determination of the work force required to carry it out. Thereafter, the proper organization, the implementation of an effective equipment management program, use of the best information, and the application of the best technology take over to ensure profitability.

CHAPTER 7

Improving Work Force Productivity

LABOR CONTROL

Modern maintenance strategies will apply new technologies and quality information along with modern management techniques. But who will ensure that the personnel who do the work are productive? The only way that maintenance can control the cost of their work is the efficiency with which they install materials. This makes the control of labor a vital maintenance function. Productivity measurements remain the most effective way of verifying the quality of labor control. But very few maintenance departments check productivity. This chapter discusses the value of improving productivity, along with a proven method of measuring it easily and effectively.

DEFINING PRODUCTIVITY

Productivity is the measure of the quality of labor control. Labor control consists of two elements:

- Labor utilization—The percentage of time the work force is available for productive work during a scheduled working period or shift.
- Labor productivity—The percentage of time that the work force or individual workers are at the work site with tools performing productive work.

These must be improved one at a time. Poor labor utilization is a factor of bad work habits, inadequate supervision, and poor attitudes. It is corrected primarily with better discipline through quality leadership. Poor labor productivity is linked to poor preparation for work, inadequate coordination, or simply not knowing what to do. It is corrected with a well-defined and executed maintenance program.

FACTORS INFLUENCING UTILIZATION

The major factors that reduce labor utilization are as follows:

- Late starts—Reporting to the work site after the scheduled starting time
- Unsanctioned breaks—Unauthorized absences from the work site
- Sanctioned breaks—Authorized absences from the work site for lunch or coffee breaks
- In-plant travel—Travel away from the work site to obtain parts or tools
- Early quits—Leaving the work site before the end of the shift

Thirty percent of a typical 8-hour shift of 480 minutes is used on nonproductive activities (Table 7-1). Of the 480 minutes available for productive work during each shift, personnel are available to do productive work for only 337 minutes, or 70% of the total time. This means that there is already a significant handicap on improving productivity because of poor work habits, resulting in nonproductive time that left only 5.7 hours of the 8 hour shift for productive work (Figure 7-1).

Failure to control nonproductive activities can result in a significant loss of valuable time during which personnel could be effectively utilized. Therefore, the constructive enforcement of authorized lunch periods, coffee breaks, and end-of-shift cleanup periods are logical first steps toward improving productivity.

TABLE 7-1 Typical summary of nonproductive time

Nonproductive Time	Minutes	Percent
Late starts	11	2.3
Unsanctioned breaks	44	9.2
In-plant travel	28	5.8
Early quits	10	2.1
Sanctioned breaks	50	10.4
Total	143	29.4

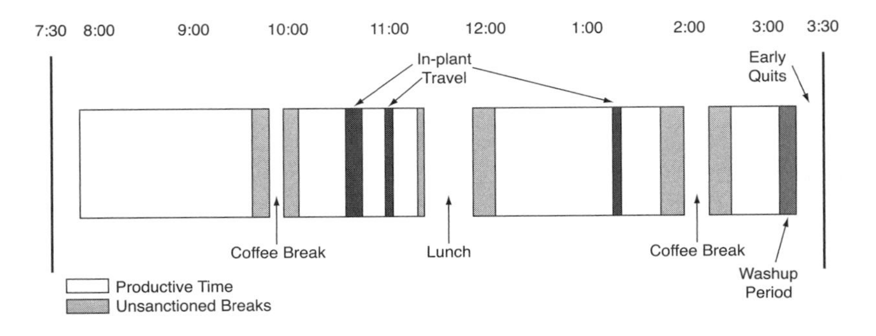

FIGURE 7-1 Only 5.7 hours of each 8-hour shift are available for productive work

However, in an era featuring the empowerment of employees, education about productivity is useful, but employees must want to improve productivity. It cannot be legislated. Therefore, their involvement in the overall process of improving productivity from the initial measurements through the improvement stage will solidify their commitment to help.

EXAMINING PRODUCTIVITY

The work site can be an area around the equipment where personnel can be easily called on to do work or where they are engaged in related work, like positioning an overhead crane. Tools can include a person's eyes as he inspects or a blueprint as he checks an assembly, as well as a wrench as he applies it. The common term "wrench time" can be misleading. Productive work is actual engagement in the job. It is the amount of productive time spent performing productive work.

FACTORS REDUCING PRODUCTIVITY

Among the factors that limit productivity at the work site are

- Interruptions resulting from excessive emergency repairs;
- Work that is not effectively planned;
- Scheduling that is poorly coordinated;
- Work site activities that are poorly controlled;
- Labor use that is not allocated properly;
- Equipment shutdown that is poorly coordinated;
- Materials and tools that are not assembled before jobs start;
- Use of support equipment that is not effectively coordinated;
- Operations personnel who fail to follow the schedule; and
- Operators who have not cleaned or prepared work areas.

The most significant aspects of poor productivity are the financial losses the plant suffers when work is not done, or is done poorly only to be redone as equipment fails again prematurely. Consider the following scenario:

> A crew of 10 paid $17 per hour would be paid $353,600 annually. But at 40% productivity, they would only perform 8,320 hours of work. This has an actual work value of only $141,400. Thus, the $353,600 annual labor investment loses $212,200 through poor productivity. Few companies would be profitable if they did nothing about improving productivity. See Figure 7-2.

The potential for labor savings as a result of better productivity can be a powerful incentive for an industrial plant to start the improvement process. Greater productivity means that the same work can be performed with fewer

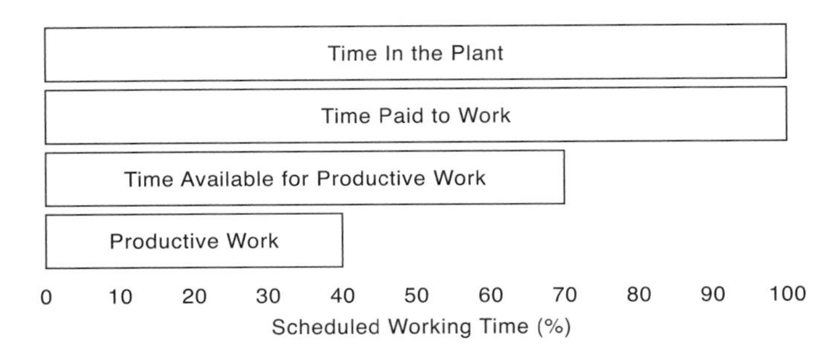

FIGURE 7-2 Wide variances exist in the amount of time maintenance crews are actually in the plant and paid versus the time they are available for work and the time they actually spend performing productive work

people. This translates into an increase in the work load capacity without having to add personnel. This is a very attractive option for any plant. In the following example, when the productivity of 30 employees is raised from 30% to 40%, it is the equivalent of adding 7.5 employees working at 40% productivity:

30 employees at 30% productivity = 16,200 man-hours per year
[30 (1,800 man-hours/employee/yr) (0.30)]

30 employees at 40% productivity = 21,600 man-hours per year
[30 (1,800 man-hours/employee/yr) (0.40)]

Added labor by increasing productivity:
21,600 – 16,200/0.40 (1,800) = 5,400/720 = 7.5 employees

The cost savings possible by performing work more efficiently is equally compelling. In the next example, the cost of performing 1 hour of work dropped from $83.86 to $66.70 when productivity was improved from 33% to 42%:

cost of 1 hour of work = labor + materials/hours

assume: 72 employees at $14/hr plus $5,100 in benefits

cost per employee = $34,220
labor = 72 × 34,200 = $2,463,840
labor = 42% of total; total = 5,866,286
total – labor = material = $3,402,446

At 33% productivity:

$$\frac{2{,}463{,}840}{(1{,}800 \times 0.33)72} + \frac{3{,}402{,}446}{(1{,}800)72} = \mathbf{\$83.86/hr}$$

At 42% productivity:

$$\frac{2{,}463{,}840}{(1{,}800 \times 0.42)72} + \frac{3{,}402{,}446}{(1{,}800)72} = \mathbf{\$66.70/hr}$$

ASSESSING PRODUCTIVITY

Workers' attitudes, lack of training, or a difficult labor environment are often contributing factors to poor productivity. But more than likely, an inadequate maintenance program that has been poorly defined, inadequately explained, and seldom followed is at fault. It follows that the first step in improving productivity is to measure it and find out how good or bad it is. The perceived negative reactions of the work force in being measured must be set aside. If productivity is not measured, the improvement process never gets started.

TRADITIONAL PRODUCTIVITY MEASUREMENTS

Random sampling is the starting point for measuring productivity. Random (work) sampling is a statistical technique of data gathering based on the laws of probability. It assumes that a sufficiently large number of observations, made at random intervals, will provide a reasonably reliable picture of what is actually occurring most of the time. The greater the number of observations there are, the more reliable the results will be. Random sampling is a method by which maintenance work force productivity can be measured.

Random (work) sampling determines the percentages of time that the group is working, idle, traveling, performing clerical functions, or waiting. It is not a measure of work force performance, nor is it a stopwatch effort. It does not measure pace against which incentive rates will be established. Nor is it an effort to pinpoint the efforts or deficiencies of individuals. Instead, it measures the effectiveness of the control of labor. Sampling is normally carried out by a third party. But third-party involvement often increases the very suspicions that are sought to avoid. Maintenance workers will see themselves as victims not benefactors of the measurement process. Productivity measurements should involve craft personnel directly. This involvement creates an opportunity for education and a corresponding reduction in resistance to the measuring process.

AVOIDING MEASUREMENTS

Overcoming resistance to measuring productivity is the first hurdle. Knowing the reasons frequently offered for not checking productivity is an advantage in neutralizing them. Consider the following reasons:

- Measurements will be misunderstood.
- The work force will suspect layoffs.
- Personnel will feel threatened.
- Measurements are not accurate.
- We don't trust industrial engineers.
- We don't have time to check productivity.

Anyone confronted with such resistance might conclude that it is counterproductive to measure productivity. If so, nothing happens. Productivity, if poor, remains that way. Worse yet, productivity that could easily be improved is not. Current levels are not known, and the factors that inhibit productivity are never identified.

INVOLVING CRAFT PERSONNEL

Why not involve craftsmen in checking productivity? Some might counter, "But isn't that like 'putting the monkeys in charge of the bananas'?" If a measure is made of the delays that inhibit productivity, it is possible to deduce the level of productivity. Craftsmen can measure these delays easily and accurately as they go about their regular work. All maintenance managers have to do is tell them how. The formula:

$$100\% - \text{nonproductive time} - \text{delays} = \text{productivity}$$

The factors that cause a loss of productivity can be clearly identified so that corrective actions can be initiated. As a result, work is made easier for craftsmen, and they become the benefactors of better productivity and not the perceived victims of poorly executed measurements. They are directly involved in the process from start to finish. It can also be argued that the craftsman is the best source of information on the actual maintenance work done. Therefore, craftsmen are in an excellent position to accurately and objectively report on delays they experience in getting their work done.

An industrial engineer can effectively measure about five productivity-related factors: waiting, traveling, clerical work, idleness, and working. But in this age of emphasis on employee involvement in which teams and business units have become increasingly popular, it makes sense to get craft personnel involved. They have inherited the responsibility for work control. Why not equip them with the means to determine how they can improve their own effectiveness? When properly trained and oriented, craftsmen can measure most elements causing delays as they perform work:

- Identifying parts
- Obtaining parts
- Waiting because there is no work available
- Incorrect instructions
- Tools are broken, not available, or not enough
- Drawings are not up to date or cannot be found
- Waiting for another craft
- Waiting for equipment to be shut down
- Waiting for lifting equipment
- Excessive clerical tasks

- Using archaic repair techniques
- More or better training needed
- Union or labor-related problems

In this era of desired employee involvement, the best weapon to reduce resistance is participation. Why not use it? If used successfully, the plant can deduce productivity and concurrently reduce delays.

MEASURING DELAYS

Random sampling is used to measure delays. However, it is applied by the craftsmen themselves. Random sampling does not mean that every maintenance craftsman need be involved in the measurements. However, all craftsman should be aware of how the delay measurements are to be made, why, and what they will reveal.

- Step 1. Ensure that there is agreement that the improvement of worker productivity is a common objective of management (to utilize labor as effectively as possible) and of the work force (to be able to carry out work with minimum delays and quality work).
- Step 2. Bring the crafts group into the process early so that they will have no reservations about a "hidden management agenda" in which they might suspect they are being manipulated.
- Step 3. As the time to start the initial measurements approaches, get the politics out of the way by explaining what is to be done, why, how, and what will happen next. Craft personnel must be comfortable with the "proposal" at this stage. Meet with the work force and their supervisors to discuss the reasons for improving productivity.
- State the objective of reducing downtime, not reducing the work force. Explain what will be done with the results. Describe how data on measurements will be used and show how decisions will be made to reduce factors inhibiting productivity.
- Step 4. Explain the technical aspects of how the measurements will be made. Make craftsmen feel that they helped develop the process. Cause them to accept ownership of the process. These points might be discussed:
  - Identify the frequency of measurements (e.g., every 6 months).
  - Establish the duration of each measurement period (about 2 weeks).
  - Explain that only 10% of personnel need to participate.
  - Demonstrate how personnel collect data.
  - Explain that personnel are not identified by name.

- Show how the data will be summarized.
- Reach agreement on how data will be analyzed.
- Show how the improvement plan will be developed.
- Determine how results will be shared plantwide.
- Schedule the next measurement.

- Step 5. Provide a simple method of collecting the data and welcome constructive comments. Let craftsmen develop the reasons for the delays to get them solidly involved. Carry out several practice runs to make certain everyone understands the measurement process. Then discuss the resulting data collected from a hypothetical point of view. Ensure that every question is answered and every suspicion satisfied. Assure them of the total support of management and supervision. A possible format used by craftsmen for collecting delay data is illustrated in Figure 7-3.

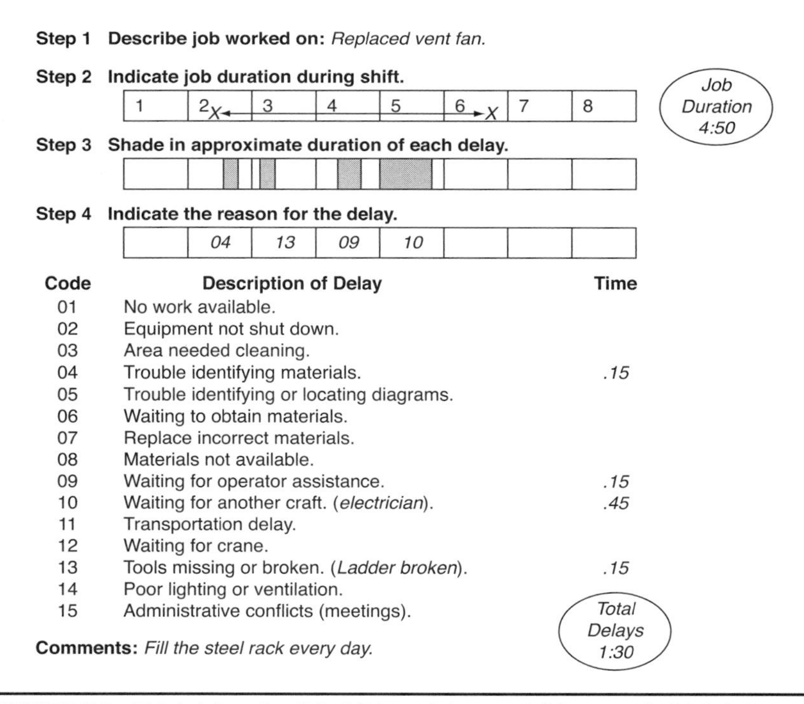

FIGURE 7-3 Total delays for this 4.5-hour job were 1.3 hours, of which 0.45 hours were spent waiting for an electrician. Note the useful comment: "Fill the steel rack every day."

- Step 6. When all of the data sheets for the measurement period are analyzed, the results should be publicized immediately. All personnel will have accepted the method and they are now looking for the answer to "how did we do?" At this stage, there is seldom any question of the validity, acceptance, or accuracy of the results. Everyone is now poised to learn "what's next?" As the results are publicized, concurrent steps should be taken to establish advisory groups to review the results and make recommendations on the best corrective actions. These groups should include craft personnel to further emphasize their participation.

  Figure 7-4 illustrates how initial measurements reveal a certain level of delays. Then, after applying corrective actions, the second measurement shows fewer delays along with an increase in productivity.

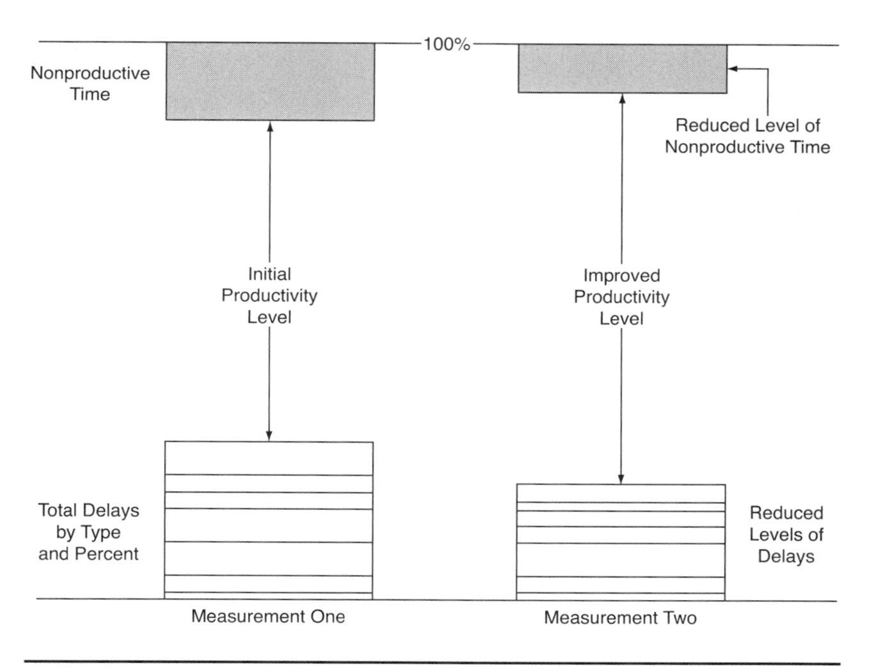

FIGURE 7-4 Measurement one establishes the "benchmark" of current delays, nonproductive time, and the equivalent level of productivity. Measurement two shows the improvement in each. The idea is to identify the delays, establish a plan for reducing them, try to reduce nonproductive time, and keep the process going. Subsequent measurements serve to monitor progress and ensure that continuous improvement is sustained.

- Step 7. After the reasons for delays are clearly established, the initial work of the crafts group has been completed. Craft personnel, new to this process, will hope that their efforts helped but they are not the managers. They will await the next steps. Management must now act. Thus, if it was discovered that more ladders are necessary or more power tools are required, management has about 24 hours to get them. Otherwise the credibility of the whole process is endangered. Obviously, longer-term improvement efforts such as realigning the procedures for parts identification cannot be done overnight. In these instances, managers must, at a minimum, acknowledge the need to correct the particular delay, state the initial steps they have taken to identify solutions, and commit to keeping plant personnel informed of progress.

If done well, the very process of productivity measurements that craft personnel once feared has now become a method by which they can improve the ease and efficiency with which they can get work done—while, incidentally, getting more needed tools, ladders, and better service. Everyone wins when productivity is improved.

SUMMARY

All plant personnel must see the value of improved productivity and become committed to getting it done. In the process, they have a favorable impact on getting better control of nonproductive time. The technique in which craft personnel measure delays, deduce productivity levels, and commit to reducing the delays is the best starting point in improving productivity.

CHAPTER 8

Understanding Preventive Maintenance

PREVENTIVE MAINTENANCE OBJECTIVES

Preventive maintenance (PM) includes all actions that avoid premature equipment failure and extend the life of equipment. These actions include routine preventive maintenance, predictive maintenance, nondestructive testing, and condition monitoring. Specifically, equipment inspection, condition monitoring, and testing are all oriented toward identifying equipment problems before equipment fails. Lubrication, cleaning, adjusting, and minor component replacement extend the life of equipment. Preventive maintenance serves as the focal point of all reliability technologies intended to isolate the root causes of failures so they can be corrected, reduced, or eliminated.

CHARACTERISTICS

PM services are routine, repetitive activities. They are routine because the same checklist or procedure is used with each repetition, and they are repetitive because the service is repeated at regular intervals. PM services are also classified as static when carried out only when equipment is shut down or dynamic when services are able to be carried out while the equipment is running. In addition, PM services are carried out at fixed intervals such as every 2 or 6 weeks or variable intervals of 250 hours or each 10,000 tons of throughput. Preventive maintenance should have a detection-orientation to find problems before they create a failure or stoppage, and utilize the time gained to plan and schedule the corrective work rather than respond to an unnecessary emergency repair.

The detection-orientation of preventive maintenance ensures that early discovery of problems will yield less serious problems, fewer failures, and more time to plan. When there is sufficient lead time between the discovery of a problem and the time when it must be repaired, there is greater assurance that the necessary work will be planned.

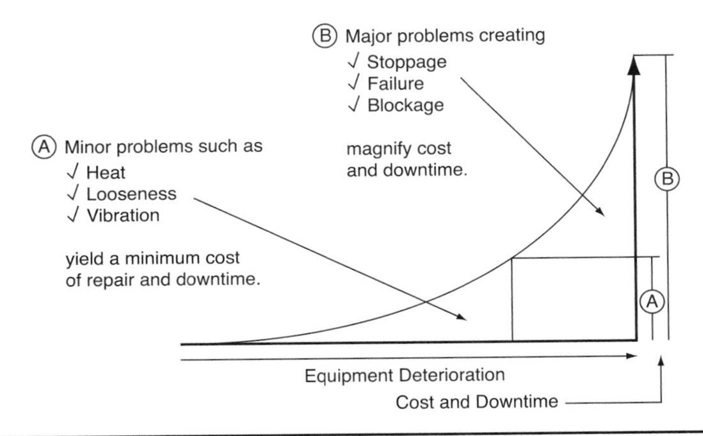

FIGURE 8-1 The level of deterioration dictates the cost of the repair and the duration of downtime

Consider the relationship of the detection-orientation of preventive maintenance also with the cost of repair and the amount of downtime to affect the repair. Essentially, problems found sooner are less serious and can be corrected at less cost and in less elapsed downtime. Conversely, problems found after the equipment has deteriorated substantially are more serious and costlier to repair (Figure 8-1). Therefore, the detection-orientation of the PM program contributes to its overall success resulting in the ability to

- Maximize equipment availability,
- Minimize downtime,
- Prolong equipment life,
- Increase equipment reliability,
- Reduce emergencies, and
- Increase the opportunity for planning work.

Of these results, the opportunity for planning more work is the most significant, as it yields better productivity, better quality work, and lower costs.

TYPES OF PREVENTIVE MAINTENANCE

There are two types of PM services: routine preventive maintenance and condition monitoring using predictive techniques such as vibration monitoring. Routine preventive maintenance includes

- Inspection,
- Lubrication,
- Cleaning,
- Adjusting and calibration,

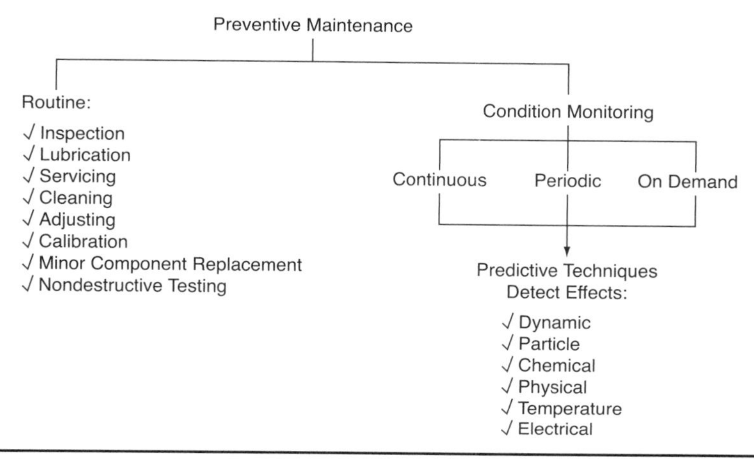

FIGURE 8-2 Condition monitoring is applied continuously, periodically, or on demand using predictive techniques oriented toward detecting specific types of defects. Note that all efforts fit with the PM activity.

- Replacement of minor components (belts, filters, etc.), and
- Nondestructive testing.

Routine PM services are carried out by maintenance craftsmen or operators as they perform visual inspections, lubrication, cleaning, adjusting and calibration, and nondestructive testing like oil sampling.

Condition monitoring uses predictive techniques to measure signals emitted from equipment. Signals are converted by specialized equipment to identify the level of equipment deterioration. These signals are then compared with signals depicting normal operation. This enables analysts to identify problems, gauge the degree of deterioration, and determine corrective actions and their timing (Figure 8-2).

Especially critical components whose failure could be catastrophic are monitored continuously. Generally, sensors located at the critical components are linked with computers set to indicate normal versus abnormal conditions.

The indication of an abnormal condition signals the operator with an alarm (called a protected system) so that immediate actions can be taken to restore the component to a normal operating condition. Continuous readings establish trends in the running condition of the equipment, and their analysis leads to the interception of a worsening condition before more serious problems arise. Typical techniques include

- Vibration monitoring,
- Infrared thermography,
- Shock-pulse diagnosis, and
- Ultrasonic testing.

IMPACT OF PREVENTIVE MAINTENANCE

The impact of preventive maintenance on the execution of a maintenance program can be best illustrated by observing the before-and-after data for

- The number of deficiencies reported,
- The time elapsed between reporting problems and equipment failure, and
- The use of labor on maintenance.

Deficiencies Reported

The importance of establishing an effective planning effort is demonstrated in Figure 8-3 as it relates the number of deficiencies found before PM program startup versus after startup. In a new PM program, deficiencies found will continue to rise until about the fifth month after the PM program has started. If those deficiencies are converted to planned work, the equipment is less likely to be troublesome at the same frequency as before the PM program was started. Planned work is done more deliberately. Thus, the frequency of repair will be reduced. As this happens, fewer deficiencies will be repeated. Conversely, if deficiencies are not converted into planned work, they will be "rediscovered" and those personnel finding them may become discouraged. If this happens, the number of deficiencies will continue to climb after the fifth month, and by about the seventh month, the whole PM program might be abandoned. Thus, as any PM program is initiated, it is important to set up effective planning

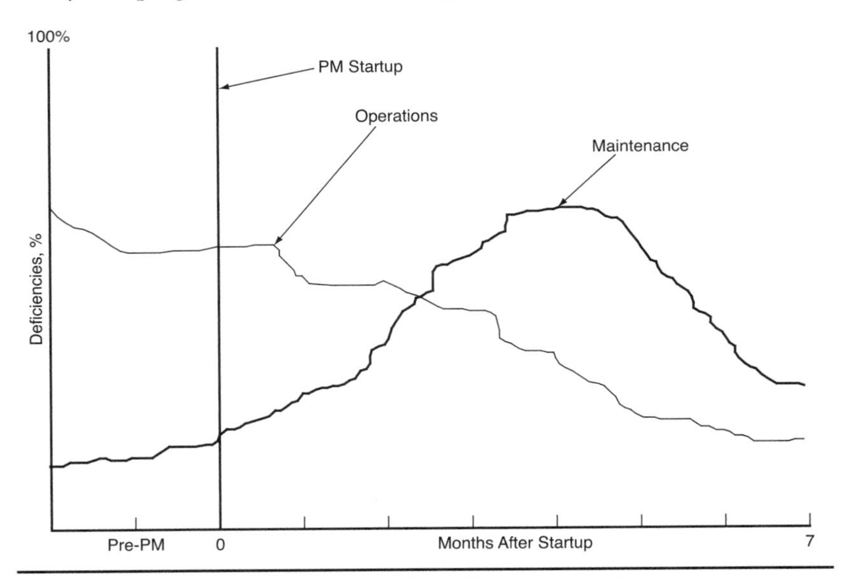

FIGURE 8-3 Deficiencies before vs. after PM program startup

to accompany it. The number of deficiencies reported by operations also falls as maintenance takes the initiative in finding problems. When the number of deficiencies uncovered by the PM program exceeds the number of problems reported by operations, it is the first sign of the success of the PM program.

Time Between Reporting and Failure

The amount of time between the discovery of the deficiency and when the job must be done is also critical for the ability to plan work. See Figure 8-4. Generally, the optimum amount of lead time for planning a job is about 1 week.

Prior to startup of a PM program, most deficiencies reported by operations are typically not known to maintenance until about 2 days before the equipment might fail. Thus, there is insufficient time to plan the jobs. However, as maintenance starts the PM program, the significance of finding problems soon enough to be able to plan them becomes important. By the fifth month, provided that the planning effort has been organized effectively, the benefit of well-planned work is apparent with less downtime per job, better productivity, and lower costs. With these improvements, it becomes much easier to convince operations of the importance of reporting problems as soon as they become aware of them. As a result, operations personnel become more aware of monitoring equipment condition and start to report problems promptly at the first sign of trouble.

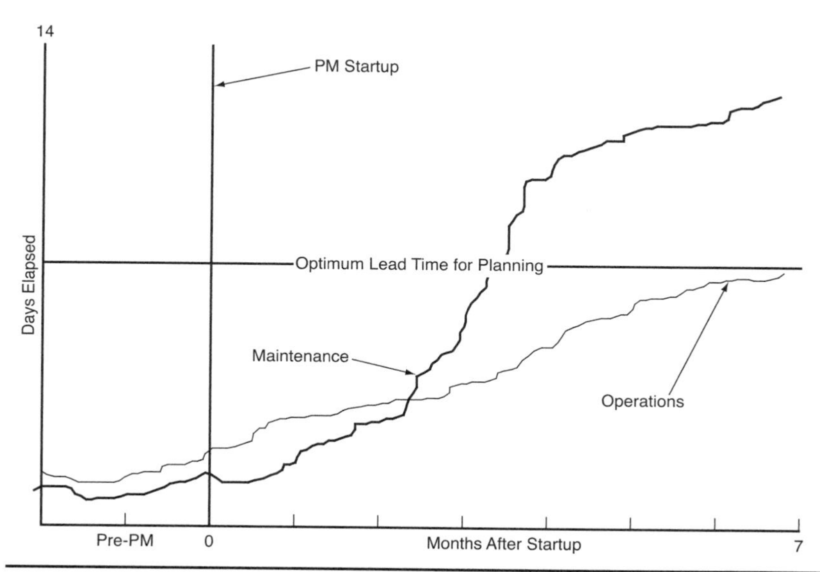

FIGURE 8-4 Within 5 months of starting the PM program, many deficiencies will be found soon enough to allow planning. The result will be an incentive to do even better preventive maintenance.

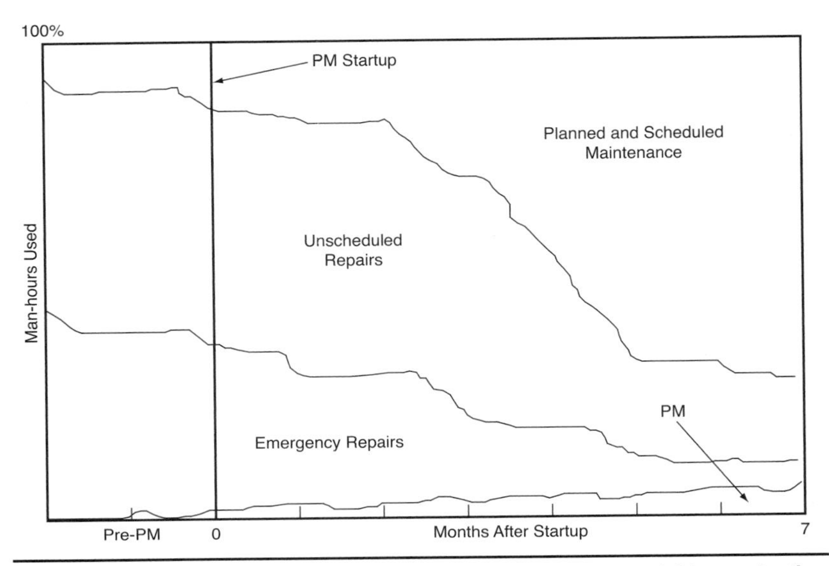

FIGURE 8-5 Optimum labor used on preventive maintenance yields a reduction in labor spent on unscheduled and emergency repairs with an increase in labor spent on planned and scheduled maintenance

Use of Labor

The PM program will also impact the way labor is used. Generally, as PM services are done more completely and on time, the amount of labor used on emergency repairs will be reduced so that more labor can be used on planned and scheduled work. See Figure 8-5.

ESTABLISHING A PM PROGRAM

Establishing the PM program consists of organizing and operating the program. After the program is in operation, experience will dictate modifications. The actions required in each step are shown in Figure 8-6.

Each key unit of equipment to be included in the program should be analyzed to determine potential equipment problem areas that could deteriorate with continuous operation. Based on this analysis, specific PM tasks can be established that will best minimize, avoid, or prevent potential problems that threaten reliable operation and performance of the equipment.

ESTABLISHING THE PM CONTROL NETWORK

Operating the PM program requires a control network to schedule, assign, and monitor services and measure program effectiveness. After PM services due are identified, a determination is made as to whether the equipment must be shut

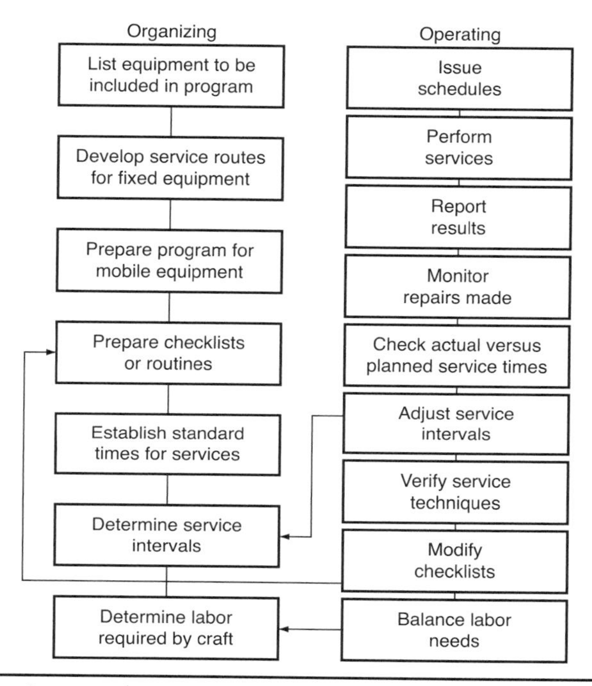

FIGURE 8-6 Step-by-step procedures for organizing a PM program

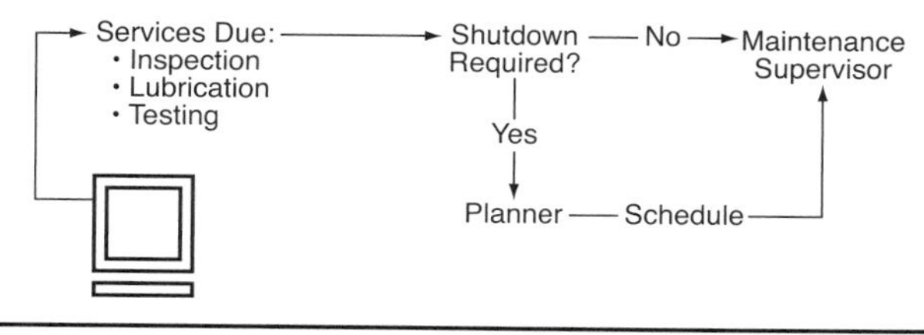

FIGURE 8-7 Control network flow to supervisor

down in order to perform the service. If so, the planner should schedule the services. If the services can be performed while the equipment is in operation, supervisors can initiate the services routinely (Figure 8-7).

After being assigned PM services, crew members conduct the services, query operators on equipment condition, list deficiencies, and advise their supervisor of the new problems uncovered or the outcome of the service. Generally, the deficiency list is written out unit by unit and component by component, and recommendations of the crew members are provided suggesting whether the repair must be made immediately (emergency repair), can wait (unscheduled repair), or might require several crafts and a multitude of materials (job planning required). See Figure 8-8.

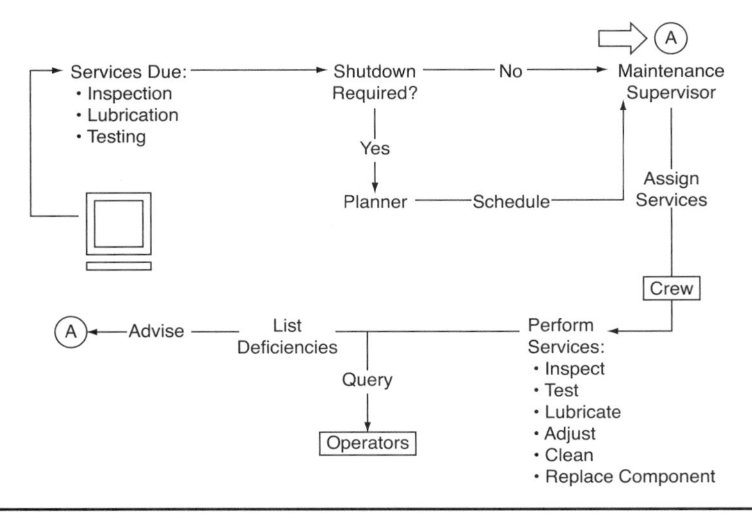

FIGURE 8-8 Control network flow to crew members

Based on the advice of crew members, the supervisor would consider each deficiency and decide on the appropriate action. The supervisor would, for example:

- Assign emergency jobs to crew members immediately;
- Enter unscheduled repairs in the computer, pending the next opportunity to shut down the equipment; and
- Take jobs that meet the criteria for planned work to the planner and discuss them with him or her.

The total picture of the PM control network is then assembled, as shown in Figure 8-9.

PM scheduling procedures within information systems identify services due based on fixed or variable intervals. The information systems contain the pertinent checklists, allowing them to be printed by the crew member at the time he or she receives the daily crew assignment.

ESTABLISHING A LUBRICATION PROGRAM

Organizing a lubrication program begins with listing the equipment to be covered. Lubrication routes should be developed that ensure all equipment in the program is serviced. Lubrication points, types of lubricants, and the methods of lubrication should be incorporated into checklists. Lubrication frequency depends on machine speed, pressures, and temperatures, as well as internal chemical decomposition or external contamination of lubricants. Manufacturers' recommended lubrication frequencies should be modified according to plant operating conditions. Then the labor required to carry out the lubrication

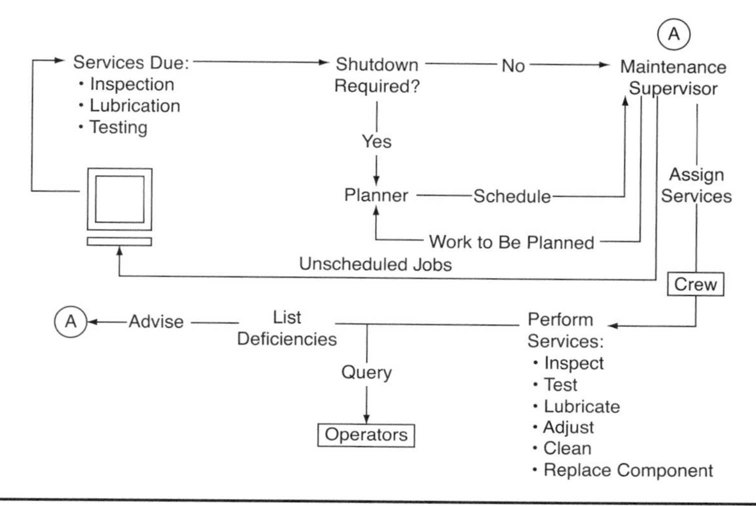

FIGURE 8-9 Entire PM control network

TABLE 8-1 Steps for organizing and operating a lubrication program

Organization	Operation
1. List equipment.	1. Publish routes.
2. Establish routes.	2. Train personnel.
3. Determine lube points.	3. Conduct services.
4. Specify lubricant.	4. Report completion.
5. Determine application.	5. Report problems.
6. Prepare checklists.	6. Analyze failures.
7. Establish service frequency.	7. Determine corrective actions.
8. Determine labor needed.	8. Make corrections.
9. Establish department responsibilities.	9. Review labor needs.
10. Set up reporting procedures.	10. Review service frequencies.
11. Make link to condition monitoring.	11. Adjust labor levels.
12. Provide for failure analysis.	12. Change frequencies.
13. Establish procedure for corrections.	13. Optimize controls.

program should be estimated. Service time for each route should include obtaining lubricants, lubricating, troubleshooting, adjusting, and reporting results (see Table 8-1).

Based on the number of routes, the service time for each, and the number of repetitions annually, the total work load can be estimated. When production personnel such as equipment operators perform lubrication activities, the labor required should include them. Reporting completion of lubrication services should follow a fixed procedure. It would include labor used as well as pro-rata charges for lubricant consumption. Lubrication failures should be reported in the same way as equipment failures, and their occurrences should be linked with an oil analysis/tribology program to ensure that root cause problems are reported and eventually eliminated.

Operating the Lubrication Program

As soon as routes, schedules, and instructions have been established, they should be documented and the training of personnel started, including operators if used. Data reported as the lubrication services are performed should be used to monitor the program and determine whether

- Allocated service times are adequate;
- Failures are reported;
- The number of failures reveal a need to alter route frequencies;
- Failures indicate wrong lubricants or poor techniques; and
- Lubrication schedules are being met, especially by operating personnel.

Control of Lubrication

Central control of the lubrication program is often provided through organizations such as maintenance engineering to ensure

- Proper coordination of oil analysis and tribology,
- Effective coordination with the overall condition-monitoring program,
- Good consumption records,
- Efficient lubricant storage,
- Use of outside technical advisors when required,
- Collection of used oils for purification or resale,
- Attention to new techniques, and
- Adequate training of personnel.

THE SUCCESSFUL PM PROGRAM

Generally, a consistent 85% compliance rate with the PM schedule will yield about 9%–11% of total maintenance labor spent on PM services. This ensures the reduction of emergency and unscheduled repairs. As this happens, 50% or more of all maintenance labor will be utilized on planned and scheduled work. Deficiencies found as a result of performing inspections, testing, or monitoring are then classified as

- Potential emergency repairs,
- Unscheduled repairs (running repairs), or
- Possible planned and scheduled maintenance.

The labor used to correct the deficiencies found should be classified as emergency, unscheduled, or planned maintenance. It should not be called "PM," which was only the effort to "find" the deficiencies. One important index of the success of the PM program is the amount of labor used and its impact on reducing emergencies while increasing labor used on planned work. Next, the compliance with the program should be measured week by week. PM compliance reporting should be included in the performance evaluation checked at each weekly scheduling meeting. If compliance is less than 85%, the reasons for noncompliance should be determined.

MAINTENANCE ENGINEERING

The objectives of maintenance engineering are to ensure the maintainability and reliability of equipment. The focal point of both maintainability and reliability is the PM program. Maintenance engineering plays a key role in the initial development of the PM program and they are among the key users of its output. In addition, it is maintenance engineering that responds to the increasing complexity of production equipment and the demand for modern technology to monitor equipment performance and conduct repairs. The ability to improve equipment performance is based on quality information. Some information is direct, like repair history, and some indirect, like field investigation. Repair history data, for example, indicates a possible failure pattern, whereas field investigation reveals the exact cause. Maintenance engineering performs a variety of services, including

- Development or modification of the PM program,
- Recommending nondestructive testing techniques,
- Development of standards for major jobs,
- Helping planners establish the scope of planned jobs, and
- Recommending standard repair techniques.

In addition, maintenance engineering monitors new equipment installations to ensure that they are operable before being placed in service. Thereafter, it ensures that the equipment is maintainable through testing and verifies that drawings, instructions, wiring diagrams, spare parts lists, and so forth are available. Testing and verification is important to avoid situations like this one:

> As a contractor withdrew his crew from an equipment installation job, the maintenance engineer pointed out that a new retaining wall made it impossible for maintenance to inspect and lubricate the equipment. He required relocation of the wall to avoid making future maintenance difficult as a result of contractor oversight.

Maintenance engineering also develops standards based on field investigation and engineering measurements. Once in use, the standards are verified by information on labor, material, and tools used on the corresponding work orders. For major repetitive jobs, similar information can be used to help develop historical standards. Labor standards can be developed by observing the number of actual man-hours by craft required to complete a specific repetitive job. Similarly, a standard bill of materials can be developed from historical material use data. With this information, maintenance engineering can confirm their engineered standards and develop new historical standards for other jobs.

Information also helps the maintenance engineer to narrow problems and identify possible solutions. Field investigation then reveals the exact cause and yields recommendations to correct the problems. Performance and equipment reliability are major areas of concern for the maintenance engineer. Of special interest is the performance of components, as exhibited in the following two situations:

> A concentrator was using six vendors to rebuild 44 different components. When they experienced a surge of component failures, they could not associate the failed components with any one vendor. Maintenance engineering developed serial-number tracking within the repair history, resulting in the identification of a vendor who had replaced only the seals but not the worn internal parts. This caused the concentrator to establish standards for component rebuilding, which in turn reduced failures.

> In another instance, purchasing was able to establish a mutual "quality control" visit between the vendor's shop and the plant's shop, emphasizing the importance of their work to equipment performance. As a result, both shops instituted internal controls that reduced quality problems almost immediately.

Because the maintenance engineer advises on both preventive maintenance and the standards for planned work, he is interested in labor utilization to verify how well personnel are carrying out PM services. If, for example, PM should represent 11% of labor use but only 3.2% is reported, some PM services are not being done. Verification of actual performance yields recommendations to improve compliance. Similarly, a check of backlog trends may reveal a need to adjust the craft composition of crews to ensure they can carry out required PM services. Maintenance engineering tasks, whether carried out by an engineer or delegated to personnel within the work force, require solid information. Excessive costs lead to the repair history, which leads to a field investigation. Suspected problems require special reports like root cause failure analysis to confirm the problem as maintenance engineering goes about its task of ensuring the reliability and maintainability of equipment. Thus, the requirement to manage equipment demands better information, and maintenance engineering is the ideal focal point to ensure its effective application.

PRINCIPAL CONDITION-MONITORING TECHNIQUES

Vibration Monitoring

Vibration monitoring uses signals generated by a machine defect as the source to indicate a deficiency. Mechanical objects vibrate in response to the pulsating force of a machine defect. As defects grow in magnitude, the vibration increases in amplitude. Changes in the intensity of the signal indicate a change in the magnitude of the defect or increase in its deterioration. Interpretation of the signal permits diagnosis of the problem. Electronic instrumentation can detect with accuracy extremely low-amplitude vibration signals. They can isolate the frequency at which the vibration is occurring. When measurements of both amplitude and frequency are established, both the magnitude of a problem as well as its probable cause can be determined. Thus, the application of vibration monitoring permits:

- Detection of impending problems,
- Isolation of conditions causing excessive wear,
- Confirmation of the nature of the problem,
- Advance planning to correct the problem, and
- Performing the repair at the best time.

Vibration monitoring is probably the most widely used condition-monitoring technique. It can be used in any industrial environment where rotating or reciprocating equipment can cause financial penalty if it fails. Vibration monitoring can reveal problems in machines involving mechanical or electrical imbalance, misalignment, or looseness. It can uncover and track the development of defects in machine components such as bearings, gears, belts, and drives.

Shock-Pulse Diagnosis (Bearings)

A useful diagnostic tool for detecting failing bearings in operating machines is shock-pulse analysis. It is practical for checks on the bearings of moving equipment, such as trucks and locomotives, and the bearings of stationary equipment. This technique is closely related to vibration monitoring and is often applied when vibration monitoring techniques are not able to be used in certain operating environments. Shock-pulse differs from conventional vibration monitoring in that it is insensitive to normal vibrations inherent in the structure of a working machine and focuses strictly on the high-frequency "shock" signal characteristic of a damaged machine element. Like other predictive techniques, shock-pulse gives ample warning of impending bearing failure, allowing maintenance to be done in a timely and orderly manner with the least disruption of machine and production time.

Instrumentation works on the principle that damaged bearings and other mechanical components generate abnormal amounts of high-frequency energy from mechanical shock and friction. A special accelerometer attached to the structure of the machine transmits a signal containing the shock signatures. The analyzer unit filters out low-frequency signal components. It integrates the resulting shock rate amplitude function, compares it with data stored in memory, assesses the mechanical condition of the bearing, and displays the result. Similar mechanisms can be inspected periodically and their shock functions compared with predetermined limits stored in memory. Output can be a simple good or bad bearing indication. Periodic inspection of a specific mechanism can also be made. A trend can be established by comparing current data with previous data. Thus, the rate of deterioration can be assessed and the expected life of the component predicted. On larger, more complex machines with numerous bearings, continuous diagnosis is used.

Infrared Thermography

Infrared (IR) thermography allows the detection of invisible thermal patterns that reveal potential equipment deficiencies. It identifies any problem that can be associated with excess temperature or lack of heat. The IR survey is conducted with equipment in normal operation and without equipment downtime. While maintenance can apply a great variety of predictive techniques, most are restricted regarding the types problems they can detect. Thermography, however, has a much wider scope of applications than most other predictive techniques. Electrical inspections, because they are so fast and cost-effective, are the principal utilization of thermography. Other maintenance problems, including mechanical problems where friction produces a thermal pattern, also reveal themselves thermally.

Thermography is a useful supplement to vibration monitoring. Tank levels, under the right conditions, can be clearly seen because of differences in heat capacity. Valves and steam traps, if working abnormally, exhibit a temperature difference that points to costly defects. Insulation or refractory breakdowns or reduced thermal resistance produces a thermal pattern showing the extent of the damage. The insulation in flat roofs, when damaged by moisture or leaks, can also be located. Looking for a "hot spot" may not be an accurate diagnostic tool; a plugged transformer fin or line to an oil cooler, for instance, will be evident because it lacks heat when heat is the normal operation. IR imagers are basically infrared cameras. They provide a snapshot of how actual temperatures compare with the temperatures that are expected, based on temperatures surrounding the equipment area being scanned. Output from a thermal imager can be recorded for later use or review. Typically, the image is recorded on conventional videotape that can be viewed on or transferred to a computer.

Several imaging systems now record directly to data disk, facilitating transfer into a computer database. The thermal image stored in the program can be printed in a report format and stored along with pertinent temperature data for trending analysis. Regardless of the type of imaging system used, data are interpreted from qualitative comparisons of the thermal patterns.

Such comparisons form the basis for most of the analysis in predictive maintenance. Whether one phase is compared to another in an electrical system or the upstream and downstream sides of a valve or trap are compared, locating a problem by its relative temperature is usually the basis of analysis. Thermography can define a problem better. The effectiveness of repairs can also be monitored. One of the best uses of IR thermography is to show maintenance how well their installations perform. It should be used to establish the baseline signature of new or modified equipment. That equipment should be inspected periodically, as it operates, to monitor deviations from the baseline. Thermal changes often precede problems. Solutions to problems that cannot be found with other test methods may be revealed with thermography.

Ultrasonic Testing

Ultrasonic inspection is used extensively on piping materials for the detection of surface and subsurface flaws and wall thickness measurement. Ultrasonic inspection methods utilize high-frequency mechanical vibrations. The fundamental principle involves a controlled and uniform beam of ultrasonic energy that is directed by a transducer into a test material. The energy will be transmitted with little loss (or attenuation) through a homogeneous material. It will be either attenuated or reflected by discontinuities or defects in the physical structure existing in a second location in the same material or in another material. The measurements of the energy change, either by attenuation or reflection, are the criteria employed.

To illustrate, steam and air leaks can be quite costly if steam is wasted through faulty valves, malfunctioning steam traps, or poorly designed systems. In a steam system with 150 lb of pressure and a production cost of $6 per 1,000 lb, a leak that is 1/32" in diameter, no larger than the tip of a ball-point pen, can cost $300 a year. Such leaks in any plant system can be detected easily with the help of ultrasonic equipment. Sound is usually caused by a vibrating body. The sound source produces a vibration that causes surrounding molecules to vibrate back and forth a small distance. As a result, each particle transmits motion to its adjoining particle away from the sound source. A sound wave has two main characteristics: tone (frequency) and loudness (amplitude). Tone is defined by the number of vibrations or cycles per second. Loudness depends on the force of the vibrating object and is expressed in decibels. An ultrasonic contact module is placed against a bearing to detect early wear or failure.

As the bearing goes from good to bad, the amplitude of the signal increases from 12 to 50 times. Ultrasonic sounds are sound vibrations above the range of human hearing. For most humans, 16 to 18 kHz is the upper limit, although a few people can hear sounds up to 21 kHz. Most operating equipment in an industrial plant emits ultrasonic signals that can warn of impending failure. The basic ultrasonic detection unit contains the electronic circuitry to translate the ultrasound into the audio range. The short waves of ultrasound have difficulty penetrating more than two media (e.g., air and metal); therefore, it is best to use the contact method to listen for leaks in bypass steam traps, valves, or behind walls.

Magnetic Particle Testing

Magnetic (wear) particle inspection is a nondestructive method for detecting cracks, seams, porosity, lack of fusion or penetration, or laminations at or near the surface in ferromagnetic materials. Because the process depends on the magnetic properties of the materials to be tested, this method of examination cannot be used on aluminum, brass, copper, bronze, magnesium, stainless steel, or other non-ferromagnetic materials. The magnetic particle inspection method utilizes magnetic fields to reveal discontinuities in materials. A magnet will attract magnetic materials only when it has ends or poles. If there are no external poles, magnetic materials will not be attracted to a magnetized material even though there are magnetic lines of force flowing through it. When a discontinuity is created in the circular ring, the effect is to establish minute magnetic poles at the edge of the discontinuities. This pushes some of the magnetic lines of force out of the metal path and creates an attraction for the formation of magnetic particles at the discontinuity. Satisfactory results may generally be obtained when the surfaces are in an as-welded, as-rolled, as-cast, or as-forged condition. However, surface preparation by grinding or machining may be necessary in some cases where surface irregularities would otherwise mask the indication of discontinuities.

Prior to magnetic particle examination, the surface should be dry and free of dirt, grease, lint, scale, welding flux, or other extraneous matter. The surface to be examined is magnetized by means of high-amperage current, either by placing prods on both sides of the area to be inspected or by wrapping the surface with a coil. The prod method is the most widely used for the inspection of piping materials and welds. With the current on, the area under investigation is dusted with finely divided iron particles to provide a light, uniform coating. A stream of air is generally used to remove particles not affected by the magnetic field. Regardless of the manner of producing the magnetic flux, the greatest sensitivity will be to linear discontinuities lying perpendicular to the lines of flux.

To ensure the most effective detection of discontinuities, each area to be examined should be tested twice with the lines of flux in one test approximately perpendicular to the line of flux in the other. If there is a crack or defect in the part, either at or near the surface, it will cause a discontinuity in the magnetic flux lines. The iron particles will collect at this point, outlining the defect. Surface defects will produce sharp, distinct patterns that are tightly held to the surface of the part. The patterns become broader and fuzzier as the defect causing indication is progressively deeper beneath the surface. Certain metallurgical discontinuities and magnetic permeability variations may produce magnetic particle indications similar to an indication from a mechanical discontinuity. Although these types of indications are termed nonrelevant, they cannot be accepted as such when they exceed the criteria of acceptance unless they are proven to be nonrelevant by additional testing.

Electrical Predictive Maintenance

Less commonly known among nonelectricians are the multitude of test instruments that are used quite routinely in electrical predictive maintenance. Many of them are used in conjunction with the more traditional PM techniques, whereas others have a singular application. Descriptions of common electrical test instruments are listed as follows:

- Circuit breaker and motor overload relay tester—Checks trip point and elapsed time of molded-case circuit breakers and overload relays.
- Fault locator—Detects opens, shorts, and grounds in cable and determines location of fault.
- GFCI tester—Determines condition of ground fault circuit interrupters (GFCIs).
- High-potential dc tester—Used for proof-testing in insulation of conductors and motors.
- Infrared sensor—Useful in detection of failing bearings and connections.
- Insulation-resistance tester, thermometer, psychrometer—Used to test and monitor insulation conditions of conductors, motors, and transformers. The thermometer and psychrometer are used for temperature and humidity corrections.
- Low-resistance tester—Checks low resistance paths of switches and contacts.
- Maximum load indicator—Useful in testing for overloads and balancing load on phases.
- Multi tester, ac and dc clamp-on ammeters, industrial analyzer, watt meter, power factor meter, high-voltage "hot-stick" voltmeter, magnetic voltage indicator, digital meter—Used for measuring power and controlling circuit voltage, current, resistance, and power; useful for troubleshooting and obtaining values for PM comparison records.
- Noise level meter—Determines sound levels of equipment.
- Oil dielectric tester—Tests oils in transformer and circuit breakers.
- Osciliographic recorder—Used for analysis of wave forms; provides a permanent record.
- Phase sequence indicator—Determines direction of motor rotation prior to startup.

- Protective relay tester—Used to set, calibrate, and test switchgear protective relays.
- Recording meter—Used for permanent record of voltage, current, power, and temperature for analytical study.
- Surge comparison testing—Measures and compares electrical surge capacities.
- Tachometer—Determines the revolutions per minute of rotating machines.
- Temperature indicating meter—Determines operating temperature of equipment.
- Transistorized stethoscope—Detects faulty rotating machine bearings and leaking valves.
- Vacuum tube multimeter, tube and transistor checker, oscilloscope, capacitance and resistance bridge, expanded scale meters—Used for electronic and solid-state circuit testing.
- Vibration monitor—Detects excessive vibration and locates imbalance on rotating components; used for dynamic balancing and alignment.

On-Board Haulage Truck Condition Monitoring

Typical of the techniques used in industry to monitor equipment performance are on-board computers used to monitor the status of haul truck, loader, or shovel-operating functions from systems like drives or engines. Monitored data are stored at intervals along with any faults as they occur. Information such as payload, cycle times, and number of loads hauled are monitored and stored during each haul cycle or working period. Data are accessible from the operator's instrument panel and can be downloaded to a computer for subsequent analysis.

Oil Analysis

It is estimated that about 30% of all engine failures are caused by metal particles contaminating the lubrication systems. Oil degradation causes residues and sludge to form, which decreases equipment efficiencies and can often destroy machinery components. As one of the most prominent nondestructive testing techniques that make up the preventive–predictive arsenal, oil analysis identifies specific wear mechanisms and contamination levels. Generally, there are two types of testing:

- Examining for metals to detect wear rates, inorganic contaminants, and inorganic additive levels
- Examining for physical properties to detect contamination from glycol, water, fuel, combustion by-products, particulates, acidity, basicity, and viscosity

Wear particle analysis is an excellent method for examining the lubricated components of a machine without requiring downtime for analysis of wear particles. Lubricating oil carries within it particles generated by a machine during operation. These particles can be electronically analyzed for relative size and concentration. When examined, the size, shape, and makeup of these particles will give valuable information on the wear mode and wear rate of a machine. This analysis can make early detection of developing problems possible, allowing maintenance to determine when the machine should be repaired before unscheduled failure. Often overlooked in the PM program is the additional factor of contaminant-inducted failure. Contaminants may be solid particles, air, moisture, and chemicals. Therefore, it is important to consider actions such as monitoring fluid cleanliness and ensuring adequate filtration to round out the oil analysis effort.

Tribology

Oil analysis, a long-term part of most maintenance programs, is traditionally carried out by laboratories with focus on internal combustion engines. But industrial plant equipment with a variety of deterioration causes are all treated like diesel engines. Thus, dealing with their root cause problems will require technologies that yield more information on the failure process. Therefore, oil analysis done by oil companies, which is often limited to spectrometric analysis, will not meet future industrial needs. Moreover, plants and supporting laboratories suffer from confusion between diagnostic maintenance needs and the content of lubricant analysis detail. The expectation is that many plants will obtain and use less costly but more capable diagnostic instrumentation. Lubricant monitoring has application in all phases of the PM aspect of equipment management, providing the organization moves toward a reliability approach that incorporates tribology as opposed to simple lubricant analysis. Tribology is the study of surfaces in relative motion. It focuses on all the influences in which two or more surfaces interact. It examines the total spectrum of influences that affect the surfaces, such as the nature of the surfaces, their chemistry, and the load and cycles of the equipment. In addition, the lubricant itself is also studied. Therefore, tribology is a more effective weapon in the overall PM arsenal.

Other condition-monitoring techniques include

- Gas chromatography, which monitors gases emitted from electrical insulating oils as a result of faults;
- Fluorescence spectroscopy, which detects wear metals in oils used in gasoline or diesel engines; and
- Dye testing, which detects cracks, wear, surface shrinkage, or fatigue on ferrous or nonferrous materials such as steel structures, boilers, shafts, or machined surfaces.

MEASURING SUCCESS

The success of the PM program can also be measured in tangible ways:

- More work is being planned.
- Components are lasting longer.
- Downtime is decreasing because of more planned work.
- The interval between repetitions of repairs is increasing.
- Emergency repairs are less frequent.
- Overtime is being reduced as emergencies decrease.
- There is less unscheduled work.
- More labor is being used on planned and scheduled work.
- Labor is being better controlled because of more planned work.
- Productivity of personnel is increasing.
- More total work is getting done.
- Overall costs are being reduced because jobs use less labor.
- Fewer personnel are needed for maintenance and may be shifted to other work, such as nonmaintenance project work.

But the most significant impact of a successful PM program is its potential for reducing cost through fewer emergencies and more planned work.

SUMMARY

The objectives of preventive maintenance to avoid premature equipment failure and to extend equipment life are achieved through the faithful application of routine PM services and condition monitoring. However, successful PM also depends on an effective planning effort to convert PM deficiencies into planned work. Only then will the real benefits of preventive maintenance be realized.

CHAPTER 9

Effective Planning and Scheduling

Planning is the pre-organization of selected major jobs and the identification of resources needed to ensure that the work can be carried out in the least amount of elapsed downtime, with the most effective use of resources, and completed at the lowest possible cost. Scheduling is the joint determination by operations and maintenance of the best time to complete the work as circumstances permit.

The ability of a maintenance organization to capitalize on the benefits of planning is in direct relationship to the quality of their preventive maintenance (PM) program. The detection-orientation of preventive maintenance permits the identification of problems well before equipment failure but also far enough in advance to permit planning the corrective work. Thus, the quality of PM services, especially inspection, testing, and monitoring, contributes directly to effective planning and scheduling. Although it can be said that all jobs require some degree of planning, there should be specific criteria for determining the particular jobs that require planning. If, for example, a simple job requires only one craft and parts are easily available from stock, such work requires no planning and should be handled entirely by maintenance supervisors.

PLANNING CRITERIA

Sensible criteria should be established for determining what work should be planned. The criteria will ensure that the planner is used effectively, and a common understanding between field supervisors and planning personnel will be created. Criteria should be jointly developed by planners and supervisors. Criteria should not be overly simple like "plan all jobs requiring more than 2 hours or costing in excess of $500." Criteria should be clearly identifiable, such as "a job plan must be prepared" or "a coordinated shutdown will be required." Ultimately, criteria should serve to clarify working relationships. For example, if certain jobs meet the planning criteria, the supervisor would take them to the planner and the planner would be in agreement that the work should be planned. There are two criteria stages. In the first stage, there are three circumstances under which jobs must be planned:

- The job cost and performance must be measured;
- Documentation is required to satisfy the manufacturer's warranty requirements; and
- Quality, cost, performance, safety, or environmental standards must be verified.

A second criteria stage is prescribed should none of the first-stage criteria apply. In that case if, for example, any 10 of the following 12 criteria exist, the work should be planned.

1. Work is not required for at least 1 week.
2. A job plan must be prepared.
3. Duration of the job will exceed one shift.
4. Two or more crafts are needed.
5. Work requires crafts that are not part of the regular crew.
6. Work requires two sources of materials: stock, purchased, or shop work.
7. A coordinated shutdown is needed.
8. Special equipment, like a mobile crane or power tools, is required.
9. Rigging and transportation will be needed.
10. Drawings, prints, and schematics are required.
11. Work requires contractor support.
12. The estimated cost exceeds $5,000.

The resulting criteria should then be discussed, reviewed, and tested. The end result should be a determination by field personnel on the type of work that should be sent to the planner with the planner in total agreement. Duties of the planner's job are discussed in detail in Chapter 4.

ADVANTAGES OF PLANNING

Planning creates an impact before work begins as well as during the execution of work and after it is completed. Before work begins

- The job scope is properly identified,
- A solid job plan is organized,
- Supervision knows what is expected of them,
- Crews can be advised of tasks in advance,
- Materials can be preordered, and
- Tools can be reserved for jobs.

As a result of good planning, the execution of a job will be improved by

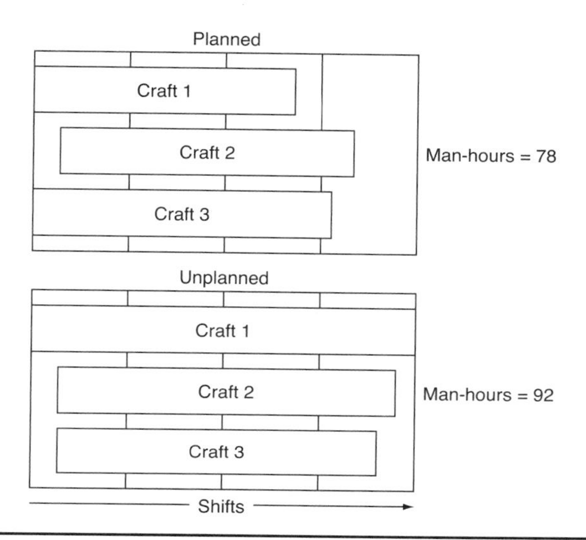

FIGURE 9-1 Upon completion of two comparable major jobs (one planned and one unplanned), it was revealed that the planned job was completed with about 15% less labor and with at least 6% less downtime

- Clearer job instructions,
- More effective supervision,
- Fewer job interruptions, and
- Better crew productivity.

The job will be completed more effectively as a result of better job preparation and execution. Typically,

- Less labor will be used,
- Each planned job will cost less in labor,
- Overall costs will be reduced,
- Jobs will be completed in less elapsed downtime,
- Better quality work will be produced, and
- There will be a longer interval before the work has to be repeated.

When work is planned, there is greater assurance that the work will be accomplished with efficient use of labor resources and in the least amount of downtime (Figure 9-1).

PLANNING STEPS

Figure 9-2 shows a typical planning, scheduling, and job execution sequence covering an elapsed period of 8 weeks. The planning steps for a particular job are shown for weeks 1–5. Then in week 6, the planned job is shown to operations,

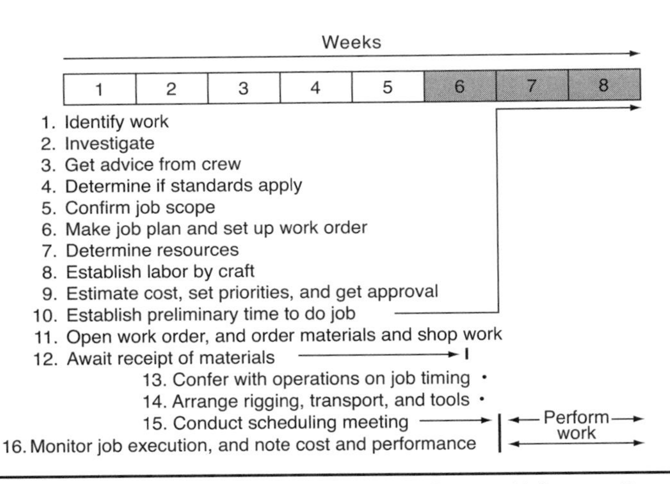

FIGURE 9-2 Logical sequence to planning, scheduling, and job execution steps

along with other planned jobs for the same future scheduling period. Next, after agreement is reached on the timing of equipment shutdown and the allocation of labor, the work is carried out during weeks 7 and 8. During this period, the planner would monitor work execution and help coordinate. As the weekly cycle repeats, the planner turns her attention to planning new major jobs. These steps can also appear in other sequences, such as a pause between planning and scheduling or between scheduling and job execution.

Planning steps shown in Figure 9-2 are discussed in more detail as follows:

1. **Identify the work.** As new work is presented, the planner verifies that it meets the criteria for being planned. She then proceeds to determine the general nature of the job and, if required, establishes a time when she can investigate the job in the field.
2. **Investigate.** The planner conducts a field investigation of the job if it has never been done before. If the job has been done before, she may consult previous work order details.
3. **Get advice from the crew.** The supervisor and crew who will perform the work are consulted to obtain their views on how the job is best accomplished.
4. **Determine if standards apply.** There are both quality standards as well as quantity standards. Quality standards describe the desired end result of work whereas quantity standards prescribe the resources required such as man-hours required. If standard task lists, material lists, and tool lists are available, they should be incorporated into the planning. A less detailed field investigation and discussion with the supervisor and crew

should still be conducted in the event that changes have been made in the equipment or its operation.

5. **Confirm the job scope.** Determine what is necessary to restore the equipment to a safe, effective, and as-designed operating condition.
6. **Make the job plan and set up the work order.** The job plan is the step-by-step approach that will be used by the maintenance crew to perform the job. Often, bar charts or critical paths are used to depict the order, sequence, and duration of each step, the crafts involved, and the completion time.
7. **Determine resources.** Resources would be identified to include labor, materials, tools, ladders, rigging, transportation, or equipment like cranes.
8. **Establish labor by craft.** For each step of the job plan, the craft labor, including the number of workers and the sequence in which they are required, would be identified. Labor needs are then balanced according to the shift on which they will be available.
9. **Estimate cost, set the job priority, and get approval.** The total estimated cost of all resources, including labor, materials from all sources, shop work, and the use of equipment, rigging, and transportation, are determined. Next, a job priority is established from the maintenance point of view. On job approval, the priority may be altered by operations. When the job is approved, it means that operations has accepted the need to perform the work, its estimated cost, the approximate timing, and its priority.
10. **Estimate preliminary time to do job.** Now the planner must tighten the timing of the job. The job cannot be started any sooner than the availability of all materials and completion of shop work. The actual condition of the equipment will reveal how long the equipment can operate before failure or trouble may occur. The job must be accomplished sometime within this "window of opportunity." If this timing is significantly different from previous estimates, planners should advise operations.
11. **Open work order and order materials and shop work.** Materials from stock, purchased materials, and shop work such as fabrication, machining, and assembly requirements are identified. Opening the work is the authorization to spend operating funds for the job. The work order is converted into an official accounting document so that costs may be debited to it. This allows materials and shop work to be ordered.
12. **Await receipt of materials.** A period of time will pass before materials, especially direct charge materials, and shop work will be available. During this period, other jobs are being planned. The planner should

monitor progress in assembling materials by observing the purchase order tracking system and work order status reports for shop work completion. Generally, stock materials would not be ordered until other materials are on hand. However, if there is concern that they may not be available, stock materials should be reserved.

13. **Confer with operations on job timing.** Just prior to the scheduling meeting with operations, the planner would assemble the preliminary plan for all jobs proposed for the following week and review them, first with maintenance, and then, informally, with operations schedulers. This will provide a basis for the preliminary allocation of labor based on when equipment could be made available.
14. **Arrange for rigging, transport, and tools.** Rigging, transportation, and tools should be arranged for in advance to better ensure that the supporting departments will be able to meet the requirement when the work is scheduled. They also appreciate the advance notification.
15. **Conduct the scheduling meeting.** The weekly scheduling meeting with operations is the opportunity for maintenance to present its plan for accomplishing major jobs and key PM services for the week following. These designated jobs may also include deferred maintenance as well as jobs that were not completed during the previous week. Based on the production plan, operations will consider the needs for equipment against the requirement to perform maintenance at the recommended times. Necessary negotiations are then carried out by the principal managers with the planner acting as an advisor. The meeting should conclude with an approved schedule. Thereafter, operations would agree to make equipment available according to the schedule, and maintenance would make its resources available to accomplish the work. The objective is to accomplish the work with the least interruption to operations and the best use of maintenance resources. During the week when the scheduled jobs are being done, daily coordination meetings between operations and maintenance are held to make adjustments in the schedule should delays be encountered.
16. **Monitor job execution and note cost and performance.** The planner should monitor the progress of each job during the week and, on the completion of each job, note its cost and performance.

PLANNING TOOLS

Maintenance planners utilize the information system to manage the major jobs they are planning from inception to completion. The work order system permits the planner to assign a work order number to each job and develop the job identification, resources required, cost estimates, and so forth.

In addition, planners can append other details to the work, such as task lists (what to do and how), bills of materials (permitting the ordering of direct charge materials or the reserving of stock materials), plus tool lists allowing the reservation of other commodities for the job. After the work order is approved and opened, it becomes an official accounting document. At this time, materials, shop work, or labor can be charged against the work order. After the work has been initiated, the accumulation of material costs, shop work, and labor use permit the planner to use a work order status report to observe the actual versus estimated accumulation of labor hours, labor cost, and material costs against the job. Upon completion of the job, the cost and performance can be measured. Similarly, planners can note the completion of all jobs on the schedule, including PM services, and provide management reports on schedule compliance. In addition, planners would observe changes in the backlog based on completion of work.

Planners are also interested in how labor is used. If, for example, preventive maintenance is effective, the amount of labor used on unscheduled and emergency work will decrease while the labor used on planned and scheduled work will increase. In addition, planners will monitor cost reports to help identify costly equipment and then track the repair history of high-cost maintenance equipment to help spot the chronic, repetitive problems or failure trends. From this information, they will recommend the development of new planned work.

SCHEDULING IS A JOINT ACTIVITY

A principal objective of effective scheduling is to perform work when it interferes least with operations while making the best use of maintenance resources. Maintenance recommends specific major tasks unit by unit together with suggested times. Operations responds with counterproposals based on the time they believe they can make equipment available and still meet their production goals. In a mining operation, for example, mobile equipment availability for maintenance is negotiated against the production or mine plan, which requires a certain "mix" of mobile equipment.

In a plant's fixed-equipment environment, usually a long-term shutdown schedule is provided to guide maintenance as they organize major tasks. At the conclusion of each scheduling meeting, there should be an approved schedule. All planned major jobs should be included in the proposed weekly plan, along with (static) PM services that require an equipment shutdown in order to accomplish the work.

CONDUCT OF MEETINGS

The principal supervisors of operations and maintenance should be present at the meeting so that binding decisions can be made. Operations might state

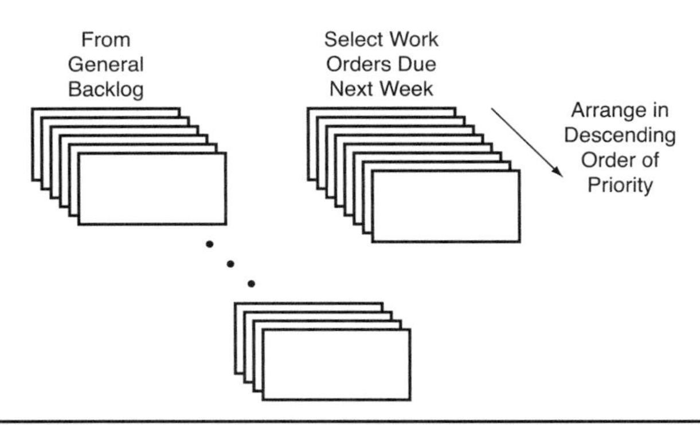

FIGURE 9-3 The typical work order system can select those work orders due the next week based on future timing assigned to each during the planning phase. Planners then arrange jobs in a descending order of priority to facilitate verification of materials availability, completion of shop work, and the ability to allocate labor to jobs.

the conditions under which it can commit to making equipment available for maintenance work. Recommendations are then made by maintenance as to which equipment they need to maintain or service, when, and for how long. Maintenance may also state the risk if equipment is not made available. Meetings are usually conducted toward the end of the week with jobs to be conducted the following week. During the meeting, it is helpful to check schedule compliance and performance for the current week. Schedule compliance is the percent of jobs completed versus those scheduled. Generally, 85% is the target. Performance information includes statistics such as the reduction in the backlog or the use of labor on emergency versus preventive maintenance or scheduled work. The purpose of checking is to note progress and ensure that corrective actions are taken. Attendance at the weekly scheduling meeting should include maintenance, operations, plant engineering, material control, and safety personnel. The plant manager should attend from time to time to judge the effectiveness and harmony with which the meeting is conducted. He or she should also listen as performance for the previous week's schedule is discussed to observe cooperation and attitudes. Each weekly scheduling meeting should be supplemented by daily coordination meetings to make adjustments in the schedule in the event of unexpected delays.

MAINTENANCE PREPARATION

Maintenance planners start the preparation for the next scheduling meeting by selecting those jobs that should be done the following week. Candidate jobs are selected for further consideration as the preliminary plan for the following week is assembled (Figure 9-3).

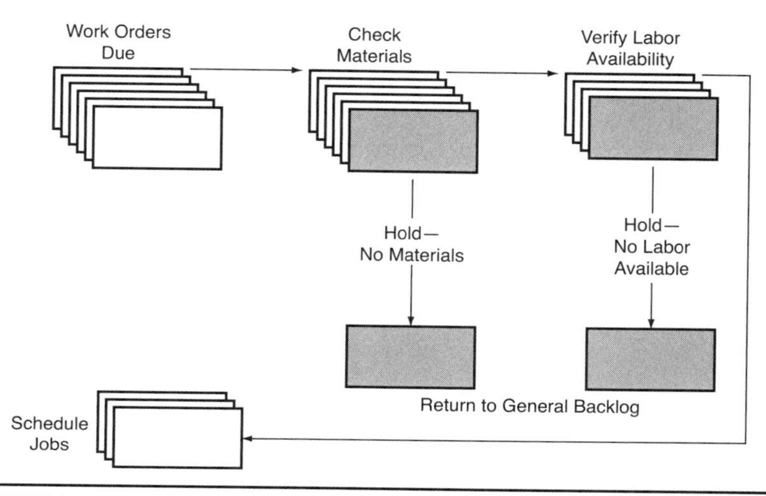

FIGURE 9-4 Materials availability and the ability to allocate labor are verified before placing work orders on the preliminary plan for internal maintenance review

The preliminary plan goes through several steps to eliminate jobs that cannot be supported. Planners verify that materials from purchasing and stock will be available. Completion of shop work is verified. Then a check is made to ensure that the labor resources can be made available (Figure 9-4).

SCHEDULING PHASES

Successful scheduling requires that maintenance assemble a realistic plan and negotiate it with operations to secure their approval of the resulting schedule. Subsequent to schedule execution compliance and performance should be verified. The scheduling process is divided into five phases:

1. Getting ready
   - Prepare preliminary plan
   - Determine completeness of planning
   - Verify availability of materials
   - Verify completion of shop work
   - Propose shutdown times
   - Discuss within maintenance
   - Verify work force
2. Conduct scheduling meeting
   - Negotiate with operations
   - Confirm priorities

  - Obtain schedule approval
  - Allocate labor
3. Advise key personnel
   - Distribute the approved schedule
   - Explain the tasks
   - Confirm shutdown times
   - Confirm use of equipment
   - Confirm delivery of materials
4. Schedule compliance and performance
   - Check schedule compliance
   - Check performance indices
5. Followup during schedule execution

Phase 1—Getting Ready

During this phase of the scheduling process, maintenance assembles a preliminary plan consisting of jobs they will recommend to be carried out for operations during the week following. Considerable detailed preparation is necessary to ensure that the plan is realistic, practical, includes essential major work, and that all aspects are ready to be carried out should operations concur. The detailed preparation steps include:

- Prepare a preliminary plan. The planner uses the work order system to help him or her select the candidate work orders for the preliminary plan describing the major jobs that will be recommended to operations at the next joint scheduling meeting. This step is usually accomplished in the early part of each week because the scheduling meeting is often conducted later in the week. This plan becomes a schedule only after operations approves it at the scheduling meeting.
- Determine the completeness of planning. After the planner has assembled the candidate jobs for the following week into a proposed plan, the elements of each candidate work order are checked by the planner to ensure they can be carried out if operations agrees with the recommendations of maintenance that the work should be done.
- Verify the availability of materials. No job can be scheduled unless the materials that the maintenance crews will install are available. Therefore, the planner should verify that the necessary materials are now on hand or will be before the work is scheduled. Primary attention should be given to direct charge materials being shipped to the plant. These are the materials over which the planner has the least control. Hopefully the planner allowed enough lead time during the planning phase to ensure that the supplier can deliver on time. Stock materials

are less critical. Generally, a check should be made to verify that stock materials will be available. If there is some doubt concerning the availability of stock materials, the planner should reserve the materials in advance. This would require the warehouse to reserve certain quantities of specific parts just as if they had been issued. However, because the job might be deferred, it is best not to actually remove the items from the warehouse shelves until the related work order has been scheduled. Some plants provide a secured holding area in which the parts are held until needed. In other instances, warehouse personnel may deliver the materials to the job site. In each instance, stock materials are charged to the work order associated with the job.

- Verify the completion of shop work. Shop work such as machining, fabrication, or assembly requires that sufficient time be allowed to ensure that the shop work can be completed on time. The work order system is useful in permitting the planner to track the status of each work order that shops are working on.
- Propose shutdown times. To ensure that the most practical plan is presented to operations, specific shutdown times should be discussed with operations in advance. This will permit the planner to start assessing labor requirements. Because craft personnel are not equally available on all shifts, special attention must be paid to allocation of labor by shift. This requires that operations personnel provide a realistic assessment of when equipment might be available.
- Discuss within maintenance. Planners, as staff assistants, cannot determine whether the preliminary plan meets all maintenance requirements, no matter how carefully they have thought it through and prepared it. Managers who are ultimately responsible for getting the work done should review the plan to ensure it is realistic based on their knowledge of overall plant operations for the coming week. They often have unique knowledge of upcoming actions that the planner does not know about. Also, in the process of reviewing the plan, the managers would assure themselves that the work can be done. Basic questions about the availability of materials and labor are asked during this internal review. As necessary, other interested personnel, like the plant engineer, might be included in the review.
- Verify the work force. A trial balance of labor availability by craft, shift by shift, should be made to ensure that the necessary work force will be available if the plan is approved by operations.

Phase 2—Conduct Scheduling Meeting

With completion of the steps leading up to the scheduling meeting, the scheduling meeting is held to obtain approval of the plan. With agreement of all involved personnel, the plan becomes an approved schedule. The objective of the scheduling meeting is to reach joint operations–maintenance agreement on the work to be done, the equipment on which it will be done, and the timing of the work. Schedule approval requires operations to make equipment available and maintenance to commit their resources to perform the work.

Decision makers should conduct the meeting. Because the proposed schedule must be negotiated within the production plan or a shutdown schedule, the most logical chairperson of the meeting is the operations manager. Similarly, the maintenance manager should present the maintenance recommendations, in the form of jobs to be done and their proposed timing. Although planners are present at the meeting, their role is to help adjust the plan in the event of unexpected changes in equipment availability. Because the planner is intimately familiar with these details, he or she is the best person to make the adjustments. When the maintenance manager presents the schedule, attendees recognize him or her as responsible for getting the work done. The earlier opportunity to review the plan generates confidence of a solid plan. Attendance should include material control and safety personnel, as required. The plant manager might attend periodically to help judge the effectiveness and harmony with which the meeting is conducted. Detailed issues during the meeting involve the following:

- Negotiate with operations. The proposed jobs are weighed against what operational circumstances will permit.
- Confirm the priorities. As the negotiations progress, priorities may change. If units originally thought to be available are now unavailable, the timing of the proposed work must be changed. The planner's detailed knowledge of resources and their availability allows this step to be carried out quickly and correctly.
- Obtain schedule approval. With successful negotiations, agreement is reached and the plan is converted into an approved schedule. There is now a "contract" between operations and maintenance to jointly carry it out.
- Allocate the labor. With the approved schedule, the planner can now allocate labor and verify that it will be available.

Phase 3—Advise Key Personnel

The planner can now notify maintenance supervisors of the work that is to be done the following week as well as its timing. Similarly, operations supervisors will be notified so they can take steps to shut down or deliver designated equipment at the proper times. This step is especially important in helping field personnel to anticipate what is coming up and to prepare for it. After the scheduling meeting, the planner should accomplish the following steps:

- Distribute the approved schedule. Distribution is important so that all concerned parties are aware of the major jobs that are to be done and their timing.
- Explain the tasks. Explanations should be given to individual maintenance supervisors so they can begin preparations and brief the crews.
- Confirm the shutdown times. Confirmation of shutdown times must be made with individual operations supervisors to ensure that there are no misunderstandings and that areas are cleaned up or equipment otherwise prepared for the start of work.
- Confirm the use of equipment. The use of mobile equipment such as mobile cranes must be confirmed to ensure they are available at the prescribed times. Service departments should be provided with information about the exact time that their support will be required.
- Confirm the delivery of materials. Confirm the on-site delivery of materials or ensure that materials are made available from designated holding areas. Timely delivery of materials adds to more efficient job execution.

Phase 4—Schedule Compliance and Performance

The conclusion of the scheduling meeting provides an opportunity to verify compliance and performance of the schedule in progress during the current week. For example:

- Check schedule compliance. Determine whether all PM services were completed and all major jobs done. If operations failed to make equipment available, find out why. Verify that material control delivered the materials on time.
- Check performance indices. Find out if the backlog was reduced and determine whether there were fewer emergency repairs as a result of better preventive maintenance. Ascertain whether labor was used more productively as a result of the planned work. Operations may also be interested in looking at the forecast of major jobs coming up in the next few weeks.

Phase 5—Follow up During Schedule Execution

As the new week's work is being performed, planners should monitor progress of ongoing work, coordinate the use of equipment between jobs, coordinate material delivery to ensure it gets to the right place on time, and participate in daily coordination meetings where adjustments are made in the timing of events to accommodate unexpected delays.

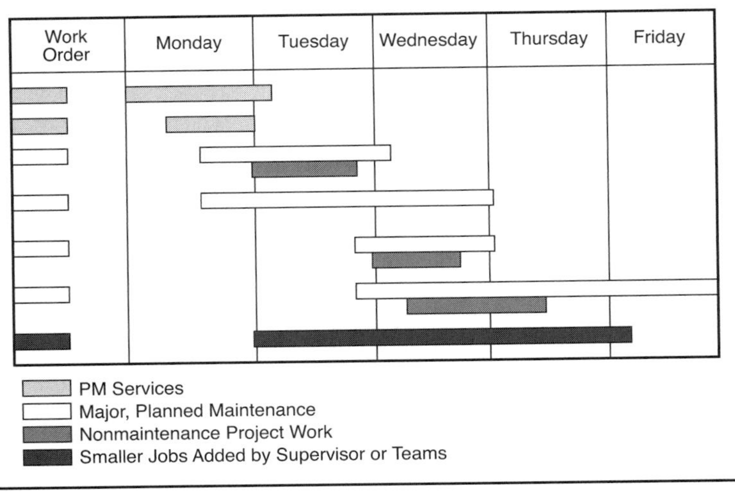

FIGURE 9-5 Sample weekly schedule

SCHEDULING PM SERVICES

Modern work order systems permit PM services due to be easily identified. Supervisors simply specify which week and the system responds with all services due by type of service, route, unit of equipment, and checklists. When services are assigned to crew members, they can print the checklists and initiate services. PM services should be given precedence in the weekly schedule, as they are the principal means of generating new planned work. Generally, static PM services that require equipment to be shut down to perform the work should appear first on the weekly schedule. PM services would be followed by scheduled major maintenance jobs. In accordance with local plant polices, nonmaintenance project work would be listed next on the schedule. When the approved schedule is presented to maintenance supervisors, they would then add smaller jobs to be done on the same units of equipment being shut down to perform scheduled work. Then crew assignments for each shift would be made directly from the schedule day by day. See Figure 9-5.

FORECASTING MAJOR COMPONENT REPLACEMENTS

A major segment of planned and scheduled maintenance includes the replacement of major components such as engines, drive lines, pumps, and conveyor drive motors. Because these components are a periodic activity repeated at the end of a specific elapsed period, the planning process can be made more efficient. Since component replacements have been performed previously, planning is simplified because each such job has

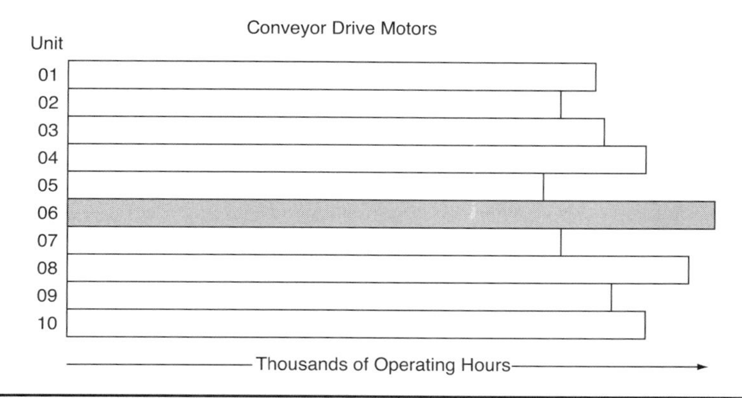

FIGURE 9-6 The MTBF establishes the timing of component replacements

- A standard task list (explaining what is to be done and how),
- A standard bill of materials (to help identify and order materials),
- A tool list (permitting them to be reserved for the job), and
- A specific labor requirement.

The task of controlling component replacements is called forecasting. Forecasting is the identification of the most likely future time when components may have to be replaced. It is important to recognize that forecasting must not be an automatic action in which, for example, a component is simply replaced after the accumulation of a certain number of operating hours or the passage of time. This type of "guesswork" will either "bankrupt" maintenance or cause maintenance craftsmen to question why "perfectly good components" are being replaced prematurely. Traditionally, the frequency at which major components are to be replaced is initially determined by establishing a mean time before failure (MTBF). This is derived from repair history, specifically, the historical life-span of similar components (Figure 9-6).

After the preliminary interval has been placed on the forecast, it can be used to help identify the possible future time when the component may have to be replaced. The initial uncertainty of the forecasted intervals must be compensated for by verifying the actual component condition before scheduling the replacement. Thus, the exact timing of the component replacement must be determined by the condition established during the latest PM inspection or condition monitoring. This step ensures the accuracy of the forecast. In time, the forecasted intervals will be confirmed and the integrity of the program established. It is important to acknowledge that the real benefit of forecasting derives from simplification of planned work, enabling more jobs to be planned. Immediate benefits include reduction in downtime for each job, lower costs,

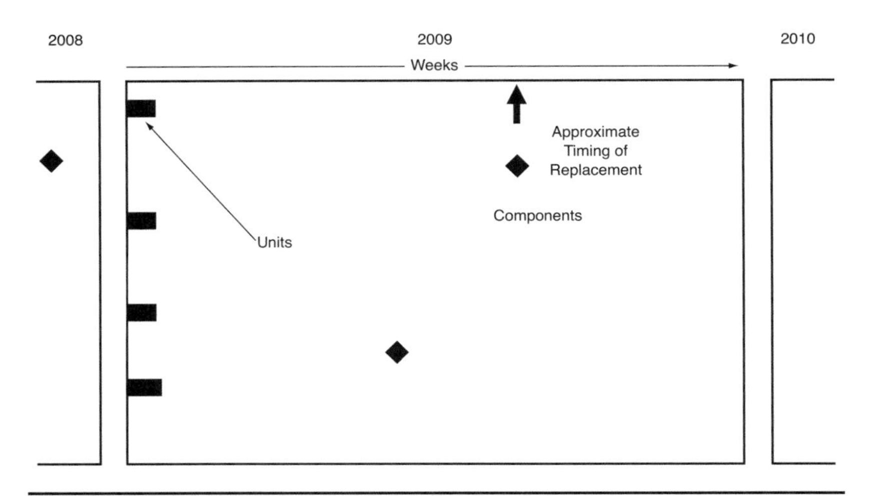

FIGURE 9-7 The forecast identifies the approximate timing when key components might be replaced

and ultimately higher quality of work and less troublesome equipment. The forecast gives maintenance a "big picture" of most major events that not only help them prepare to do the work but also simplify the planning steps associated with each job (see Figure 9-7).

Each forecasted event includes standards on how to perform the job, as well as the materials and tools required. The purpose of these standards is to ensure that the job is performed consistently with each repetition. In addition, standards provide for better planning. Jobs do not have to be investigated individually, and materials are identified in advance. The use of standards such as those shown in Figures 9-8 through 9-10 simplifies the planning and converts the event from a planned and scheduled event to a much simpler rescheduled event.

Repair history data facilitates the identification of component replacement intervals. Forecasting also links the work order to inventory control and purchasing. In addition, the forecast helps to identify the labor required to carry out each component replacement. These events also provide a picture of man-hour requirements for the component replacement program.

Most component replacements require materials from purchasing and warehousing. Both benefit from the forecasting procedure. Purchasing can anticipate the number of components to be rebuilt annually and arrange to keep the right number flowing through the rebuild "pipeline." The warehouse can ensure that the right number of associated spare parts are on hand and pull together "packages" corresponding with forecasted tasks. In addition, arrangements can be mode for on-site parts delivery and reserving of special tools needed.

Standard Task List—117
Task Description: Rebuild (RBL) Fire Water Pump (diesel, centrifugal)
Location: Process Water Circuit

1. Shut off water to and from pump and lock out valves.
2. Shut off electricity to motor and lock out motor.
3. Disengage coupling halves. Refer to drawing of Peerless Type AE.
4. Remove nuts from gland bolts (17b) and remove packing glands (17) from the shaft (6). The packing gland halves are separable.
5. Remove all nuts or cap screws from the upper casing (lb) and from the bearing caps (41 & 43). Match mark bearing caps to lower casing (la).
6. Use jack screws on the bottom side of the lower casing split flange to separate the upper and lower casing. Turn the jack screws back below the split flange surfaces to avoid interference during reassembly.
7. Attach hoist to eyebolt in upper casing. (Eyebolt is 1/2-13 thread). Remove upper casing.
8. Place slings around the shaft near the bearing housings and lift the rotating element from the lower casing. Tap lightly on the underside of bearing housings to separate the housings from the brackets.
9. To disassemble rotating element, loosen set screws and remove the coupling half. Remove coupling key (46) and outboard deflector (40b).
10. Take out cap screws and remove bearing covers (35 and 37) and the gaskets (73b). Remove inboard bearing cover seal (47) from cover (35) only if replacement of seal is required.
11. Remove retaining ring (18a) from outboard end of shaft.
12. Remove housings (31 & 33), bearings (16 & 18), and bearing seals (169) as units with a bearing puller. Remove deflectors (40a).
13. Remove casing rings (7). This may be done by removing the coupling half.
14. Remove packing rings (13), lantern rings (29), and stuffing box bushings (62).
15. Loosen shaft sleeve set screws then loosen shaft sleeves (14 & 14a) with a spanner wrench. Sleeve (14) has right hand thread. Sleeve (14a) has left hand thread. Remove sleeves from shaft. Note - A seal between the shaft and sleeve is made with an 110-1 ring (14b) in a groove in the shaft. Do not damage O-ring
16. Remove the impeller with a press or tube and hammer. Note - The interference between the impeller hub ID and the shaft OD meets standards. Do not change.
17. Remove impeller key.
18. Clean all metal parts with a bristle brush, do not use wire brush.
19. Visually inspect all parts for damage affecting serviceability or sealing. Emphasize inspection of mating parts having motion. Replace all defective parts with new.
20. Measure impeller OD wear surface or impeller ring (8) and the ID of the casing ring. If the ID dimension minus the OD dimension multiplied by 2 is greater than .015–.019, impeller will need to be replaced or repaired.
21. Inspect impeller passage for cracks, dents, gouges or imbedded material
22. Replace casing rings (7) if grooved scored or eccentric.
23. Replace impeller rings if needed (8). The impeller must be machined to install new rings. Do not reduce hub OD when machining off old impeller rings.
24. Install new impeller rings (8) on the impeller. Shrink or press on.
25. Replace worn shaft sleeves.

Courtesy of Barrick Goldstrike Mines, Inc., Elko, Nevada.

FIGURE 9-8 Sample standard task list

Standard Task List—117 Task Description: Rebuild (RBL) Fire Water Pump (diesel, centrifugal) Location: Process Water Circuit
26. To re assemble, rotating element, coat shafts lightly with oil. 27. Place impeller key (32) in shaft key way. 28. Align impeller on shaft and install with an arbor press or brass tubular sleeve and hammer. Do not bend shaft. When assembled the impeller vanes must rotate in the proper direction, and the impeller hub must be centered on shaft journal. 29. Coat shaft sleeve "0" rings with oil and install in shaft sleeves (14 & 14a). Install shaft sleeve. Set screws in shaft sleeves. Assemble shaft sleeves to the shaft and hand tighten against the impeller. Do not tighten shaft sleeve set screws. 30. Install the stuffing box bushings (63). 31. Locate casing rings on impeller. 32. Place inboard deflectors (40a) on shaft. 33. Install bearing housing seals (169) into housings (31 & 33), insert bearings (16 & 18) into housings (31 & 33). 34. Press housings-bearing assemblies on shaft to seat bearings against shaft shoulders. 35. Install bearing retaining ring (18a) in groove against the outboard bearing. 36. Install gaskets (73b) on bearing covers. 37. Attach inboard and outboard bearing covers (35 & 37). In the assembled position, the grease drain tap must be located at bottom in a horizontal plane. 38. Install outboard deflector (40b), and coupling key (46); assemble coupling half on the shaft and tighten the set screws. 39. For pump assembly, install casing gasket (73a) to lower casing (la) with shellac. 40. Use slings around the shaft near bearings to set rotating element into lower casing. Position the casing rings (7) and both bearings so that all dowel pins engage in slots in the lower case split surface. 41. Assemble both bearing caps per match marks and tighten cap screws. 42. Adjust the shaft sleeves (14 & 14a) to center the impeller in the lower casing volute. Tighten both shaft sleeves with a spanner wrench, then tighten shaft sleeve set screws to 130 inch/pounds of torque. 43. Cover the top side of the casing gasket with a mixture of graphite and oil. Install the gland bolts (17b). Then, carefully locate the upper casing on the lower, making certain the dowel pins engage. Install cap screws and tighten, working from the center of the casing to each end. Torque to the correct value. Rotate shaft by hand to check that it turns freely. 44. Push the stuffing box bushings (63) to the rear of the stuffing boxes insert two packing rings, lantern ring, and three packing rings. Insert each ring separately and stagger the joints of successive rings 90 degrees. Insert the packing glands (17) and set the gland bolts nuts finger tight. Do not use a wrench. 45. Rotate shaft by hand to check it turns freely. 46. Replace all drain plugs. 47. Re lubricate the bearings. 48. Fill with oil. 49. Reassemble coupling. 50. Turn on all water lines previously isolated. 51. Remove lockouts from power 52. Turn on pump. After pump is turned on adjust packing as needed.

Courtesy of Barrick Goldstrike Mines, Inc., Elko, Nevada.

FIGURE 9-8 Sample standard task list (continued)

Standard Bill of Materials for Task List—117 Rebuild (RBL) Fire Water Pump (diesel, centrifugal)		
Part Number	**Description**	**Quantity Required**
2692293-076	Shaft RH	1
2693431-449	Impeller 8.93"	1
2617556-056	Bearing retainer ring	1
2667213-118	Inboard deflector	1
2667213-118	Outboard deflector	1 (same number)
269249303-116	Lantern ring	2
269249203-11	Stuffing box bushing	2
05070001	Seal bearing housing	1 (2 per set)
05070002	Seal inboard brg. cover	1
05070003	Bearing inboard	1
05070004	Bearing outboard	1
05070005	O-ring shaft sleeve	2
05070006	Sleeve shaft RH	1
05070007	Sleeve shaft LH	1
05070008	Ring packing	1 (12 per set)
05070009	Ring impeller	1
05070010	Ring casing wear	1
05070011	Gasket bearing cover	1
05070012	Gasket casing	1

Courtesy of Barrick Goldstrike Mines, Inc., Elko, Nevada.

FIGURE 9-9 Sample bill of materials

Standard Bill of Materials for Task List—117 Rebuild (RBL) Fire Water Pump (diesel, centrifugal)				
Sequence	**Task Description**	**Skill**	**No. of Workers**	**Man-hours**
1	Shut off water	ME	1	0.1
2	Lock out electric pump	ME	2	2.0
2	Disassemble pump	ME	2	6.0
3	Inspect/clean pump & parts	ME	2	4.0
4	Reassembe pump	ME	2	6.0
5	Turn on water/remove locks	ME	2	2.0
6	Adjust packing	ME	2	2.0

ME = mechanic.

Courtesy of Barrick Goldstrike Mines, Inc., Elko, Nevada.

FIGURE 9-10 Sample man-hour requirements

SUMMARY

Well-executed planning makes a major contribution to improving productivity, reducing cost and minimizing the elapsed downtime to complete jobs. But because planned work is carried out deliberately, work quality improves. Thus, the components installed will perform better and last longer. Craftsmen doing the work soon realize that work is becoming easier to do. Effective planning can make a major contribution to plant profitability. The weekly scheduling meeting is an opportunity for operations and maintenance to negotiate the shutdown of equipment for necessary PM services and major work. Work is scheduled to interfere least with operations while making the best use of maintenance resources. During this meeting, performance during the previous week can be noted and corrective actions taken. The forecasting of major component replacements is an effective way to simplify planning and allow individual planners to plan more work. In addition, the use of standards provides consistent quality work, resulting in less downtime, reduced cost, and improved equipment performance. Interdepartmental cooperation is enhanced significantly when forecasting is shared with operations, warehousing, and purchasing.

CHAPTER 10

Reliability Centered Maintenance

Reliability centered maintenance (RCM) is a proactive strategy for achieving maximum equipment reliability and extended life at the least cost. Implementation identifies specific equipment functions in their exact operating context. Then equipment performance standards are identified for each function and failures defined when performance standards are not met. Based on the consequences of failures, a maintenance program featuring condition-monitoring (CM) techniques is applied to identify potential failures (equipment is starting to fail) accurately and quickly to preclude their deterioration to functional failure levels (equipment no longer operates). Thus, equipment life is extended and the consequences of functional failures are reduced or avoided (Moubray 1996).

Modern industrial production equipment has been designed for greater reliability and increased productive capacity to meet more demanding production and quality goals. But it has also become more complex and increasingly challenging to maintain. Greater technical expertise is required to realize the potential benefits. RCM when added to equipment management is an effective strategy for achieving them.

Traditional maintenance strategies built solely on time-based actions such as inspections, servicing, major component replacements, and overhauls are proving to be inadequate. All are based on the premise that equipment reliability deteriorates with age. RCM contradicts this traditional view. By identifying the nature of equipment failures, RCM specifies a strategy that reduces the consequences of equipment failure. Consider, for example, a 190-ton haulage truck as might be used in mining. Its primary function is to move ore or waste from a loading point to a crusher or waste dump. It must carry out this function to meet a specific performance standard:

> Carry a 190-ton load up and down 12% grades at speeds of up to 30 mph during all weather conditions for periods of 24 hours, stopping only for refueling, periodic operator checks, and shift changes.

Any condition that does not meet the stated performance standard would constitute a failure, of which there are two types (Moubray 1994):

- Potential failure—An identifiable physical condition which indicates that the failure process has started.
- Functional failure—The inability of a unit of equipment or component to meet a specified performance standard.

To determine early detection of potential failures and accurately distinguish them from normal operating conditions, condition-monitoring techniques such as vibration monitoring are used. These techniques detect deteriorating equipment conditions with greater accuracy than routine physical inspections. By correctly applying condition-monitoring techniques, RCM is able to forestall both the equipment failure and the consequences of failure.

THE EQUIPMENT FAILURE PROCESS

Generally, components of any mechanical equipment are subject to wear, corrosion, and fatigue. As deterioration increases, the reliability of the equipment decreases. Unless detected and corrected, the deterioration of components increases until the equipment fails. Maintenance observes, detects, and corrects the process of failure. They do this by inspecting and servicing at prescribed intervals. Then, anticipating the age at which components are likely to fail, they replace them or perform overhauls at predetermined times. This timing often has no bearing on actual equipment condition. Industrial maintenance, under philosophies such as time-based overhauls, has paid less attention to how components fail and the consequences of failure. Experience with RCM shows an unwarranted assumption that components wear out and become less reliable as operating age increases. Thus, the standard operating procedure in industrial maintenance has been to try to restore equipment to an as-new condition by periodically replacing components or overhauling the unit. In doing so, maintenance has tended to overlook the failure process itself and the question of what constitutes a failure (Moubray 1994). In turn, this omission has led to a maintenance strategy of avoiding downtime and production loss rather than one based on a wider range of consequences should equipment fail. Because these consequences impact everything from environment to economics, they demand attention. Condition monitoring, which can obtain precise evidence that a failure is occurring, is a far better means of avoiding failures and their consequences than are time-based inspections, component replacements, and overhauls. Therefore, RCM implementation requires an explicit understanding of the failure process and the dedicated application of condition-monitoring techniques to preclude failures and avoid their consequences (Moubray 1994).

UNDERSTANDING EQUIPMENT FAILURES

Successful RCM implementation requires understanding of failures and how they occur. Failures must be considered primarily in the context of the equipment user. Thus, a deviation from the normal operating condition that is unsatisfactory to the equipment user would constitute a failure. Consider this example:

> An engine that runs unevenly at high speed is considered a failure to a race car driver but perhaps not to the everyday driver. Similarly, a worn tire that holds air will allow driving in the city, but its lack of adequate tread is a tire failure to the winter driver on a snowy mountain pass.

A failure is an unsatisfactory condition. But the difference between unsatisfactory and satisfactory depends on the kind of equipment and the operating environment. To illustrate,

> A haulage truck engine with low compression would preclude the truck from climbing a 12% grade with a 190-ton load. To maintenance, the truck runs. But to the pit operations supervisor who must meet production targets, the truck has failed.
>
> Similarly, a gear case that leaks oil has failed in the view of the safety department because the oil spillage can endanger personnel working nearby. However, the same gear case is only a failure to the maintenance engineer when it leaks excessive oil and threatens to fail. Yet to the production supervisor, the gear case is a failure only when it ceases to drive the conveyor system.

Failures, as in the gear case situation, suggest that the unsatisfactory condition can range from a physical indication that it may stop to complete stoppage. From the maintenance view, failures are classified as potential failures (an indication that the failure process has started) or functional failures (do not perform designated functions to the performance standard specified). Potential failures include

- Vibration signaling the onset of possible bearing failure,
- Cracks indicating the start of fatigue in metal casings or frames,
- Hot spots showing the deterioration on smelter furnace casings,
- Metal particles in gear box oil indicating possible gear failure, and
- Excessive wear on tire treads (see Figure 10-1).

The most effective RCM application requires that the functional failures be defined and the means of detecting potential failures with condition-monitoring techniques specified.

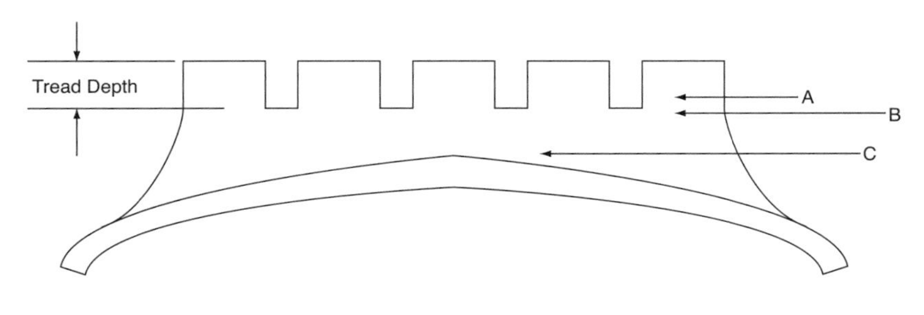

Source: Adapted from Nowlan and Heap 1978.

FIGURE 10-1 A potential failure is suggested at wear level A. At wear level B, the tire is smooth and further threatens failure. But if wear continues to level C, the loss is a functional failure.

CONDITION-MONITORING TECHNIQUES AND APPLICATIONS

Condition monitoring is built on the fact that most failures will give some type of warning that they are going to occur. This warning is termed a *potential failure*, which is the physical indication that a functional failure is in the process of occurring. A functional failure means that the equipment cannot meet its specified performance standard. The condition-monitoring techniques used to detect potential failures are called "on-condition" tasks. Equipment is inspected, tested, or monitored and then left in service on condition that it continues to meet its specified performance standard. The frequency of these actions to determine whether equipment will be left in service is determined by the P–F interval, which is the interval between the identification of a potential failure (point P) and its deterioration into a functional failure (point F). Basic on-condition tasks are essentially like the human senses. But although humans can detect problems over a wide range of potential failure conditions, the P–F interval of humans is so short that they may be watching the functional failure with little ability to provide any lengthy warning period. Thus, the sensible addition of condition-monitoring techniques to an otherwise satisfactory preventive maintenance (PM) program offers the promise of a more effective detection-oriented program (Figure 10-2).

With modern condition-monitoring techniques, potential failures can be detected sooner, resulting in a longer P–F interval. This means more time is available to take the actions that will avoid functional failure and its consequences. Table 10-1 illustrates typical effects of equipment operation, the failure symptoms they exhibit, and the condition-monitoring techniques used to detect them.

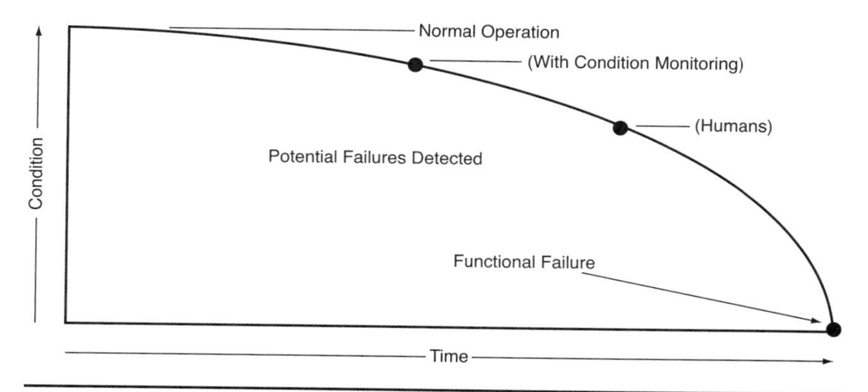

FIGURE 10-2 Modern condition-monitoring techniques allow earlier, more accurate detection of potential failures and quicker responses. This results in avoidance of more failures and extended equipment life.

TABLE 10-1 Condition-monitoring effects and symptoms vs. techniques

Effects	Failure Symptoms	Techniques
Dynamic	Abnormal energy emitted in vibration, pulses, and noise	Vibration monitoring Shock-pulse diagnosis Ultrasonic monitoring
Particle	Particles released into operating environment (oil)	Oil sampling Wear-particle analysis
Chemical	Chemicals released into operating environment (air)	Gas chromatography Fluorescence spectroscopy
Physical	Visible cracks, fractures, and dimension changes in equipment	Vibration monitoring Dye testing
Temperature	Abnormal increases in the temperature of the equipment (not the product)	Infrared thermography
Electrical	Changes in resistance, conductivity, dielectric strength, and potential	Infrared thermography Surge comparison testing

RCM IMPLEMENTATION

Implementation follows an eight-step procedure, starting with the selection of the most critical equipment and ending with a complete program incorporating RCM (see Figure 10-3):

1. **Select the most critical equipment.** As an example, assume that the most critical equipment in a particular operation is the 190-ton haulage truck fleet.
2. **Identify the functions of the most critical equipment.** What exactly do these haulage trucks do in their operating context?

   The primary function of the 190-ton haulage truck fleet is to move ore or waste from a loading point to a crusher or waste dump.

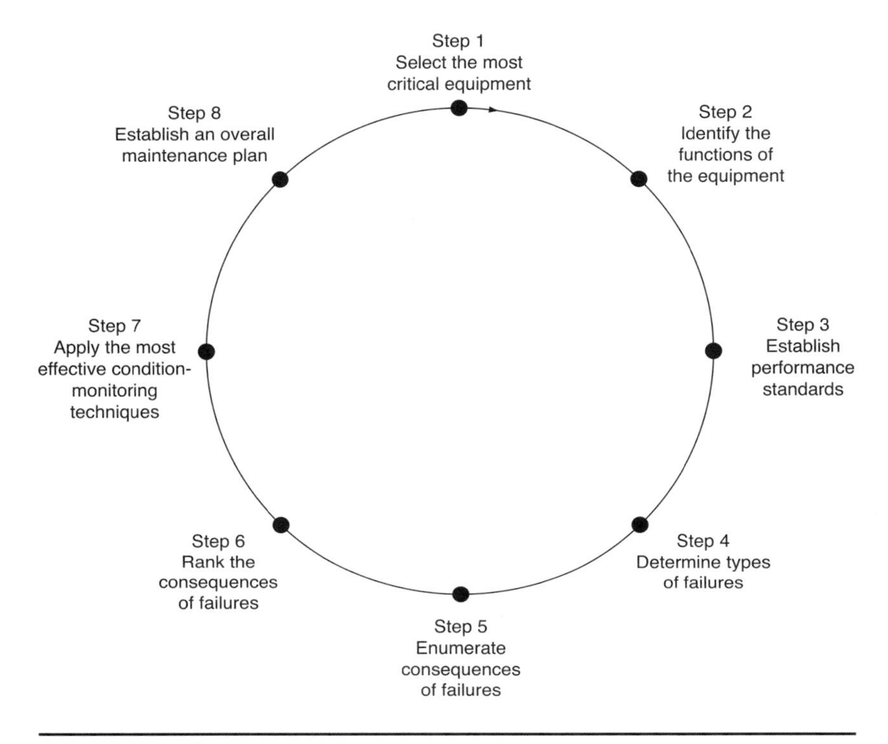

FIGURE 10-3 Steps in the RCM implementation process

3. **Establish performance standards.** How well must the haulage truck perform in the conditions under which it operates?

   Each truck, for example, must be able to carry a 190-ton load up and down 12% grades at speeds of up to 25 mph during all weather conditions for periods of 24 hours, stopping only for refueling, periodic operator checks, and shift changes.

4. **Determine the types of failures.** Any equipment condition that does not permit a haulage truck to meet the performance standard would constitute a failure.

   A potential failure is an identifiable physical condition which indicates that the failure process has started. On haulage trucks, typical potential failures might be

   - Vibration signaling the onset of transmission failure;
   - Cracks indicating the start of fatigue in the truck frame; or
   - Metal particles in engine oil, indicating possible bearing failure.

   A functional failure is the inability to meet the specified performance standard. A haulage truck experiencing the following types of failures would not be able to meet its performance standards and would sustain a functional failure:

- Hydraulic pressure insufficient to raise truck bed
- Low engine compression reducing engine power
- Electrical shorts inactivating warning devices

One must also be aware of hidden failures in which the failure is not apparent until the function is attempted by the operator and the truck fails to respond. Typically, these might include instances where

- The operator pushes the brakes pedals but the truck keeps going, or
- The operator activates the lever to raise the bed but nothing happens.

5. **Enumerate the consequences of failures.** What will the result be if a specific failure occurs?

   Consequences of failure can range from inconvenient to catastrophic. For example, a haulage truck with failed warning lights can be restored to its performance standard with little downtime as the offending fuse is found and replaced. However, a haulage truck without brakes can pick up speed, collide with another truck, damage both trucks, injure the drivers, block the road, and require extensive repairs.

   In the larger context, maintenance can affect all phases of the industrial operation. Without reliable equipment, production targets cannot be met. Without dependable equipment, product quality and customer satisfaction goals are not met. Unreliable equipment can endanger personnel, create environmental hazards, and even undermine energy efficiency. For all of these reasons, avoidance of the consequences of failure becomes a primary objective.

6. **Rank the consequences of failures.** Modern industrial production equipment has increased in complexity, multiplying the number of ways it can fail. Therefore, failure consequences must be classified to guide preventive and corrective actions. For example,

   - Safety failures endanger personnel as well as equipment.
   - Operational failures result in product loss plus the cost of repair.
   - Nonoperational failures result only in the cost of repair.

   In industry, the most important aspects are the avoidance or reduction of the consequences of safety and operational failures. Therefore, the most competent types of condition-monitoring techniques are applied to the equipment most critical to the safety of individuals and the production process.

7. **Apply the most effective condition-monitoring techniques.** To detect potential failures early and accurately distinguish them from normal operating conditions, condition-monitoring techniques such as

vibration monitoring are used. These techniques are capable of detecting deteriorating equipment conditions with much greater accuracy and reliability than can humans. These techniques also detect hidden failures that human beings would not be able to detect unless they tried a control mechanism and it did not respond (e.g., lifted lever and truck bed did not raise). With the availability of more effective and reliable condition-monitoring techniques, equipment condition can be more accurately observed. This allows equipment to remain in service on condition that it continues to meet its performance standard rather than replacing the troublesome component at the first sign of potential failure. In turn, this approach yields significantly greater life from components and units.

8. **Establish an overall maintenance plan.** Based on the consequences of failure, a maintenance program featuring condition-monitoring techniques is applied to identify potential failures accurately and quickly to preclude their deterioration to functional failure levels. The most effective maintenance program is built on the preceding implementation steps. Then the condition-monitoring techniques selected are fitted into existing, competent maintenance programs to protect the 190-ton truck fleet from functional failures and their consequences.

CRITICAL EQUIPMENT

Determination of the most critical equipment should involve operations, maintenance, and engineering so that all concerned participate. A simple formula should be applied so that an initial identification of critical equipment can be quickly established. Subsequently, the listing can be changed if warranted. The initial simplicity of the procedure is intended to get beyond the first step of implementing RCM as quickly and as accurately as possible, knowing that subsequent experience may alter the listing as more experience is gained (Figure 10-4).

RELIABILITY

Reliability, the overall condition of production equipment measured by the extended life-span of internal components is one of the objectives of both RCM and total productive maintenance. (Total productive maintenance is described in more detail in Chapter 11). Efforts to achieve better equipment reliability began slowly. The run-to-failure attitude of the 1930s gave way to basic inspections by 1950. By 1968, equipment failure studies had revealed a need to seriously try to avoid both the failure of equipment and its consequences. By 1990, advanced condition-monitoring techniques made effective strategies like RCM achievable (Figure 10-5).

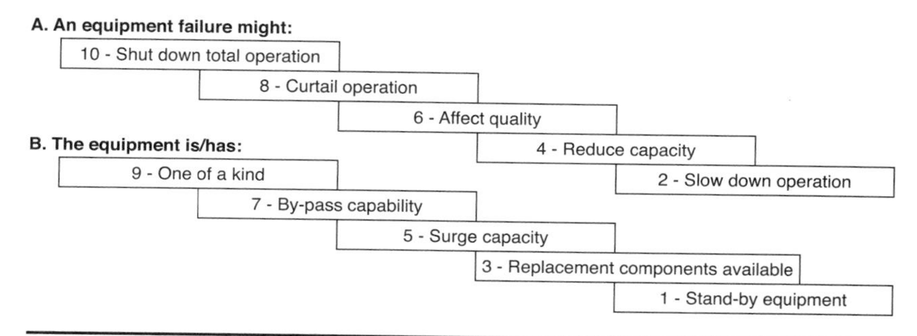

FIGURE 10-4 Operations, maintenance, and engineering collaborate to identify the conditions under which A and B might apply as illustrated. Then each unit of candidate equipment is classified under condition A and B. The product of the two classifications is the initial criticality of the equipment.

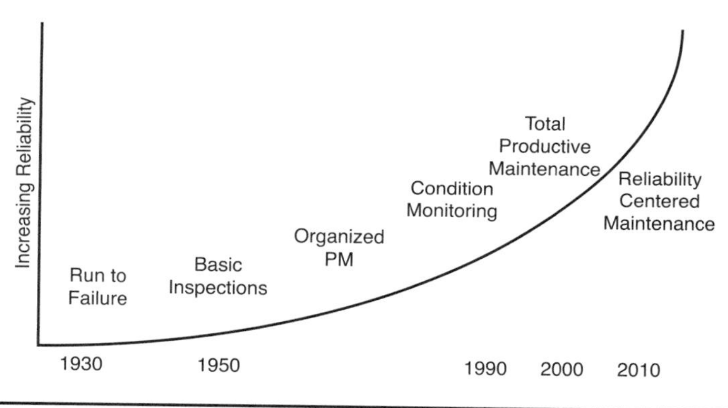

FIGURE 10-5 Reliability efforts over the years

Prior to the first comprehensive study of RCM, only two failure patterns were acknowledged: those expected to wear out as the equipment got older and those that needed breaking in and subsequently wore out with age (Figure 10-6). Limited knowledge of failure patterns led to the belief that most failures could be avoided by replacing major components or overhauling equipment at the end of a predetermined wear-out period. By 1968, additional studies had confirmed the existence of six failure patterns but with only slightly different percentile distributions (Figure 10-7).

Repair history is the primary means by which data is developed to determine the shape of failure patterns and the degree to which failures occur versus the life-span of the equipment or components being considered (Figure 10-8). Repair history data compiled from the pump impellers described in Figure 10-8 resulted in the failure pattern illustrated in Figure 10-9.

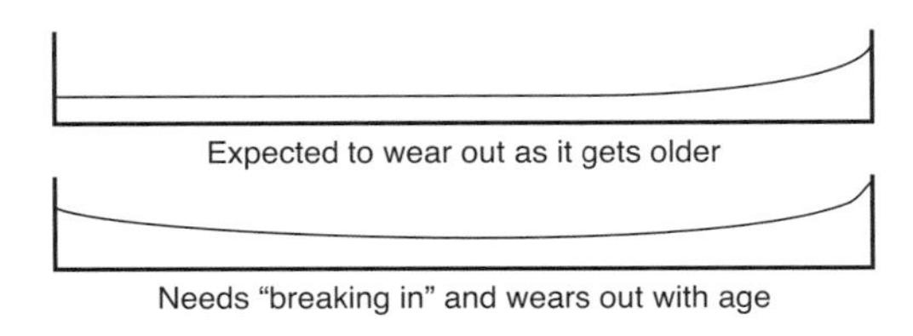

FIGURE 10-6 Initial failure patterns

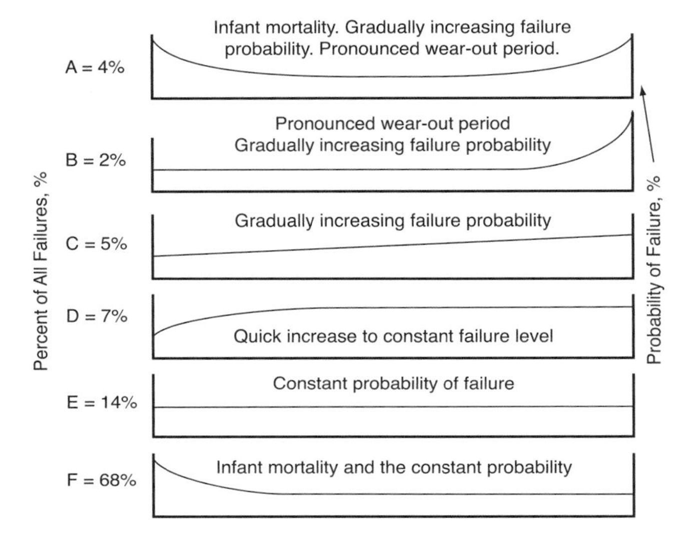

NOTE: Although Heap and Nowlan established these failure patterns in 1978 and they were attributed to aircraft, Moubray confirmed them in 1994 for a wider range of industrial equipment. Moubray's confirmation was based on 10 years of experience in the implementation of RCM in 27 different industries. In addition, the U.S. Navy further confirmed these patterns in their studies to better understand potential failures on nuclear submarines (Allen 2006).

Source: Adapted from Nowlan and Heap 1978.

FIGURE 10-7 Of the six patterns identified, 89% (D, E, and F) of failures were not related to age. Only 11% (A, B, and C) had failures related to age.

REPONSES TO SPECIFIC FAILURE PATTERNS

Each failure pattern will require corrective measures based on the nature of the incidents resulting in the failures recorded in repair history. A detection fault might require more intensive condition monitoring. An operating error might require changes in the speed at which the equipment is being operated. Material fatigue or corrosion might dictate that more attention be paid to the quality of spare parts being utilized.

Infant mortality failures, for example, suggest potential problems in maintenance, operations, or materials procurement. Maintenance, for example, may have inadequate work control procedures, personnel in need of further training,

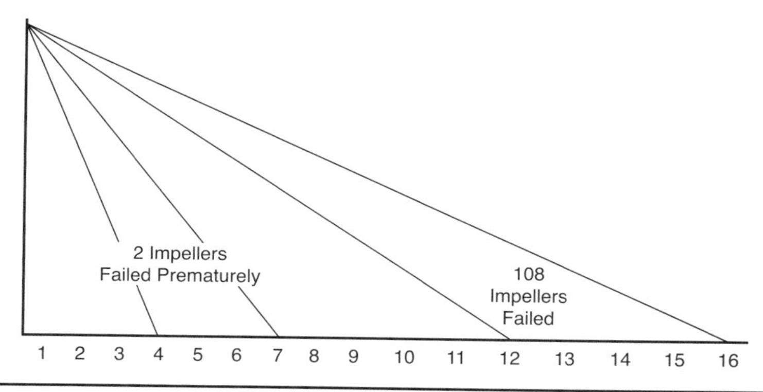

FIGURE 10-8 The repair history of 110 pump impellers were observed over a 16-month period. Two failed prematurely at months 4 and 7. The remaining 108 impellers lasted between 12 and 16 months.

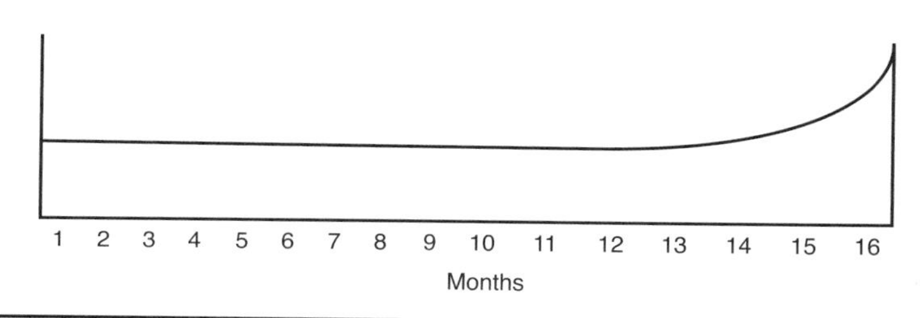

FIGURE 10-9 Failure pattern from data in Figure 10-8

and possibly poor supervision. Operations may need to review startup procedures or train operators more effectively. Materials procurement may be buying materials based solely on price or procuring from too wide a variety of vendors. However, until the root causes are determined, there are no time-based maintenance activities for these types of failures.

Random failures, on the other hand, are usually caused by some outside action that induces failures into the component. Maintenance may not have well-designed lubrication routes or may be inconsistently tightening bolts and possibly neglecting cleanliness practices. Operations may be operating equipment outside its design envelope or have inadequate quality control of incoming raw materials. Materials procurement may not have clear parts specifications. Increasing the focus on the basic PM program could eliminate many random failures.

MAINTENANCE RESPONSES

Implementation of RCM opens a new door for maintenance responses. During the period when only two wear-out failure patterns were recognized, avoidance of failure was the primary objective, and the responses were limited to

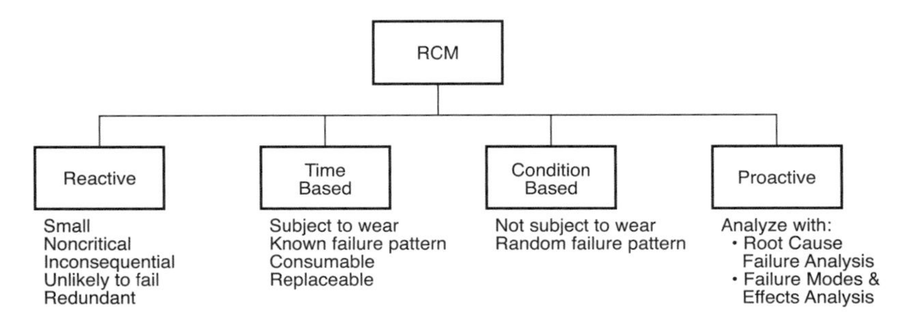

FIGURE 10-10 RCM must include all forms of responses in order to satisfy the full range of potential failure patterns as well as the ability to accurately identify underlying causes of failures

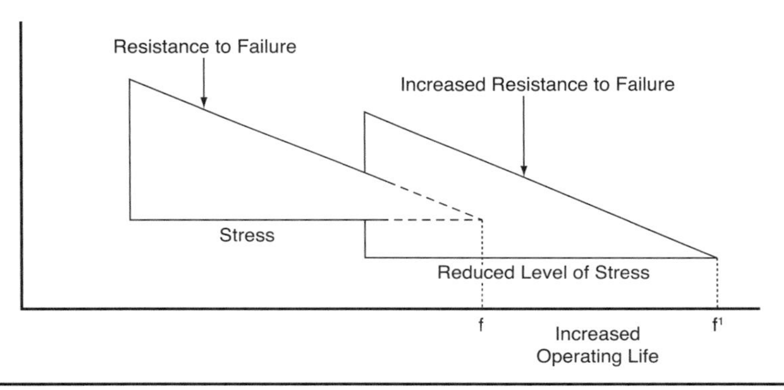

FIGURE 10-11 Maintenance can increase resistance to failure with actions like installing more rugged components. Operations can reduce stress with slower startups and lower operating speeds. As a result, the operating life of equipment could be increased from f to f^1.

time-based actions such as periodic component replacements or overhauls. With the determination of other failure patterns exhibiting random failure patterns unrelated to wear, the introduction of condition-based actions became necessary. It would protect against both premature equipment failure as well as the consequences these failures would induce. As a result, RCM included responses that are reactive, time based, condition based and proactive (Figure 10-10). Concurrently, responses will dictate maintenance actions to increase resistance to failure while reducing operational induced stress (Figure 10-11).

APPLYING THE P–F CURVE

Condition monitoring is accompanied by the ability to accurately identify the potential failure of equipment. Then, using data derived from repair history for the type of equipment under consideration, the P–F interval (the elapsed time

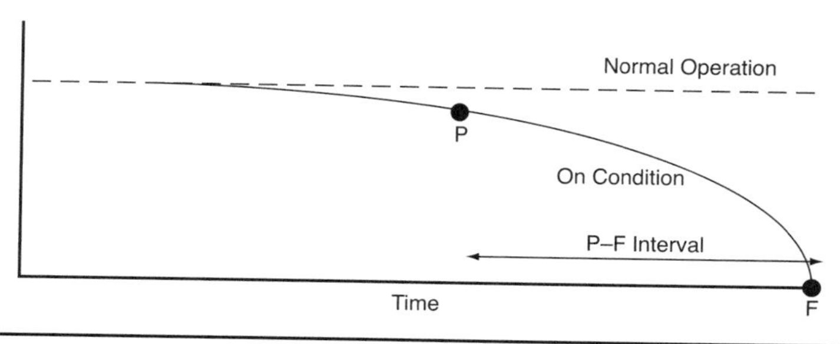

FIGURE 10-12 Applying the P–F curve

until functional failure [F] would occur if no corrective action were taken) can be established. The longer the P–F interval, the greater the time to take corrective actions. During this period, the equipment is more carefully monitored and left in operation on condition that it can continue to perform its designated function. Corrective actions taken during the on-condition period attempt to restore the equipment to its normal operation. When successful, the life-span of the equipment is extended (Figure 10-12).

UPDATING THE PM PROGRAM

RCM implementation will alter the nature of the PM program by changing the "detection" processes. Inspections conducted on time-based intervals may give way to condition-based monitoring. Some time-based services might be supplemented with condition-based monitoring. Some equipment might be subject to redesign to better meet new requirements for reliability. Other equipment, especially of a noncritical nature, might be allowed to run to failure (Figure 10-13).

Generally, if equipment displays random failure patterns at any point in its repair history, condition-based monitoring alone or in combination with time-based inspections will be more effective in reducing the instances of failure (Figure 10-14).

APPLYING REPAIR HISTORY

Repair history records the chronological listing of significant repairs made on critical equipment. It reveals the nature of chronic, repetitive problems, confirmation of failure modes, observation of failure trends, determination of failure frequencies, and the life-span of components. Repair history is a key ingredient in the determination of the adequacy of equipment monitoring as well as the principal means of determining whether corrective actions have been effective (Figure 10-15). From this information, solutions can be developed and results observed.

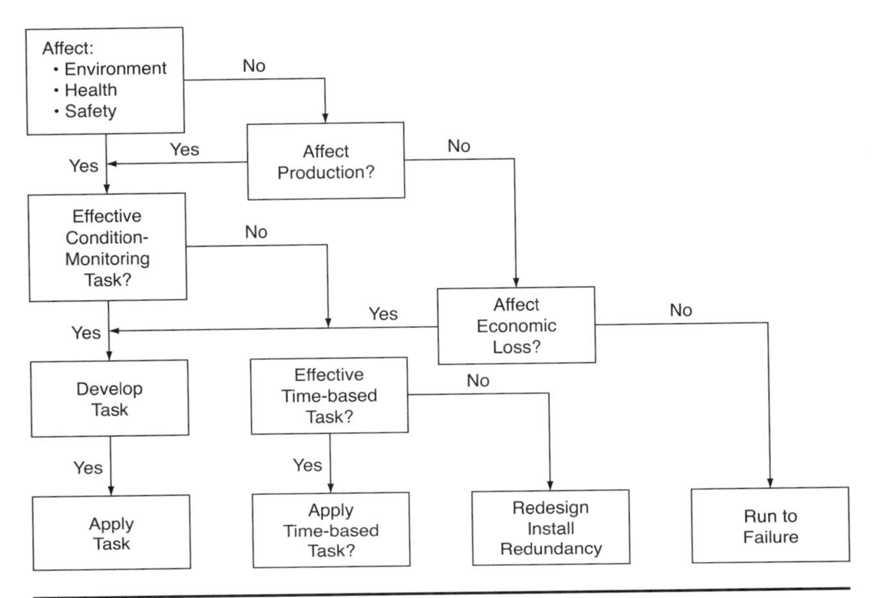

FIGURE 10-13 A simple decision tree can be used to determine which equipment will require condition-based monitoring versus time-based inspections or be allowed to run to failure

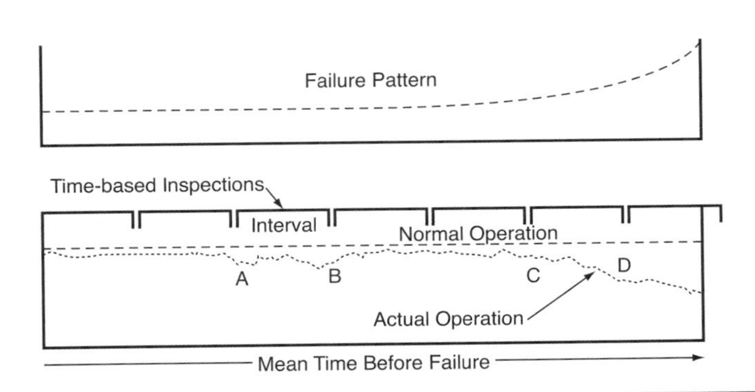

FIGURE 10-14 The application of time-based inspections would have missed equipment problems at A, B, C, and D based on the actual operation illustrated. In this instance, the best solution would have been to augment time-based inspections with the selected use of condition-based monitoring.

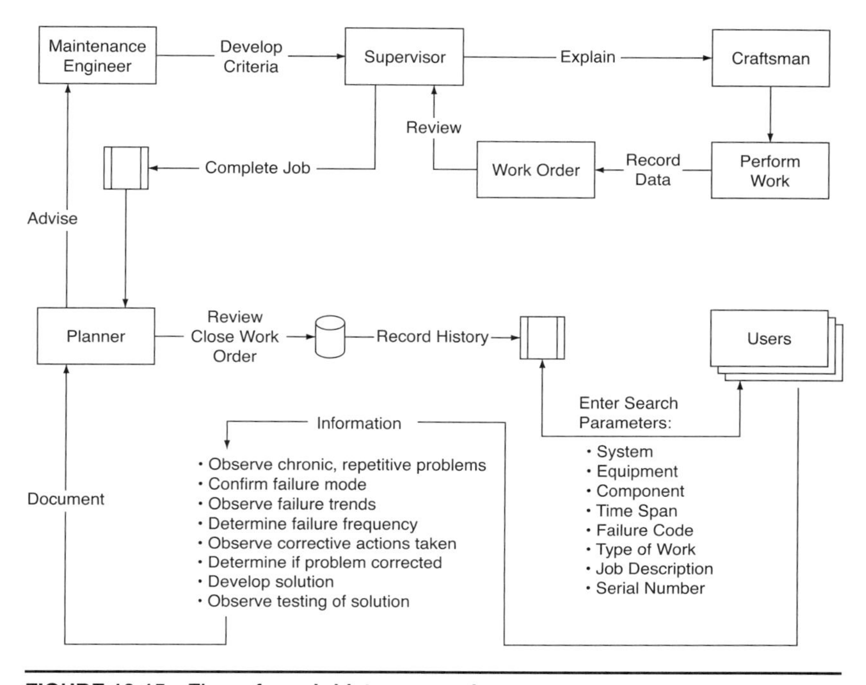

FIGURE 10-15 Flow of repair history records

ACTIONS DURING THE ON-CONDITION STATUS

After the P–F interval has been established, an on-condition status is assigned to the equipment being observed. Observation includes more vigorous monitoring to establish the underlying nature of the newly discovered potential failure. During this period, useful maintenance, operations, or material solutions are actively sought to try to restore the equipment to its normal operating condition (Figure 10-16).

ANALYTICAL TOOLS TO ACCOMPANY RCM

Reliability tools help to establish the underlying causes of failure and assess associated risks. Root cause failure analysis, failure modes and effects analysis, and risk-based inspections are among such proactive maintenance tools. Investigations of equipment failures begin with inspection of the damaged equipment. Discussions with involved personnel are followed by an examination of pertinent records. When all of the relevant data are assembled, contributing factors are assessed to arrive at recommendations (Figure 10-17).

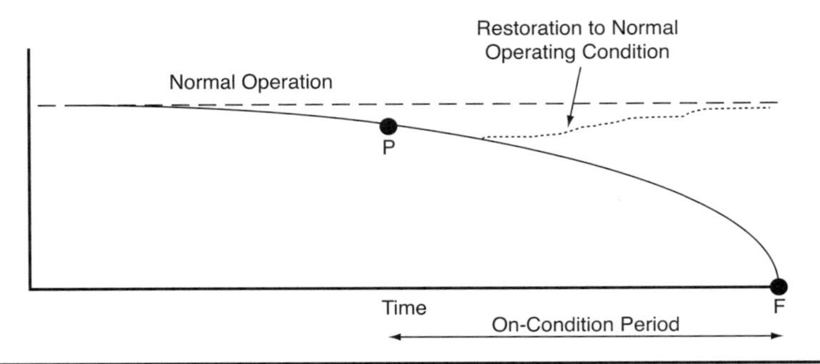

FIGURE 10-16 The P–F interval establishes a period of time during which the equipment is left in operation while corrective actions are sought to try to restore the equipment to its normal operating condition. If successful, the life-span of the unit can be extended.

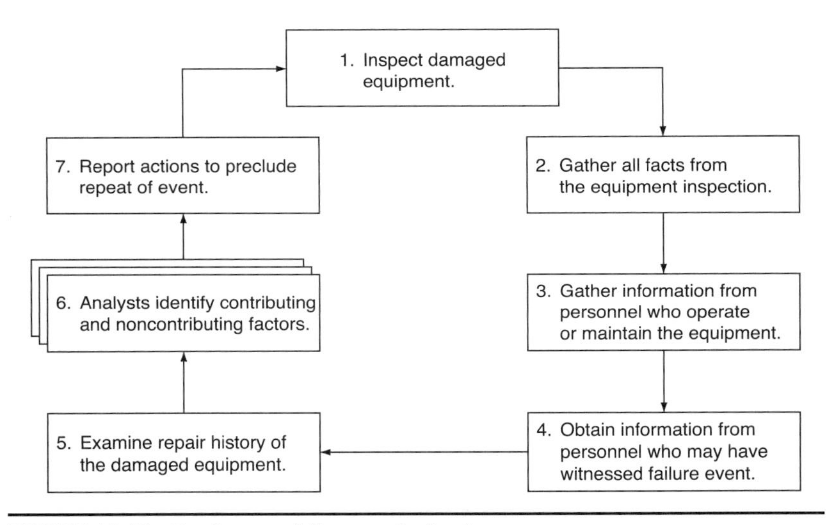

FIGURE 10-17 Root cause failure analysis steps

As with any investigation, care must be taken to preserve the objectivity of the inquiry. Consider this experiment:

> Twelve observers were asked to watch a volleyball game and count the number of times the ball was hit into the air until one team reached 21 points. While the game was in progress, a video was made and a person dressed in a gorilla outfit circulated among the players. When the game was finished, the observers were asked to report how many times the ball was hit into the air. Everyone's answers were approximately correct. When asked if they saw the gorilla, none of them did.

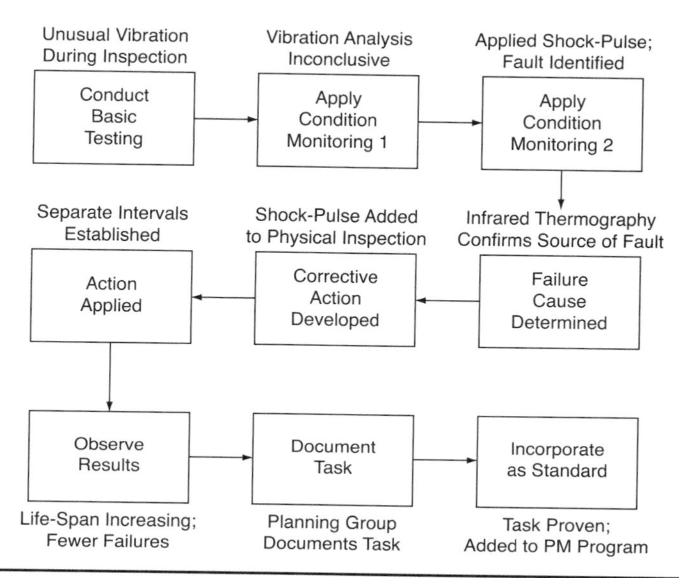

FIGURE 10-18 This illustrative RCFA includes three possible condition-monitoring solutions. The first was inconclusive. The second successfully identified the root cause. But a third CM task was administered to verify that the second had accurately found the root cause. From this investigation, a solution was proposed, tested, and documented, then converted into a standard with wider application on similar equipment.

> But when they saw the video, everyone was surprised that they had missed seeing the gorilla.

Investigations should be "evidence driven." But there can be serious deviations from this guideline. For example,

> In a classic case of departure from a root cause analysis conclusion being evidence driven, one team reached an acceptable conclusion they thought their manager would accept. Then they searched for evidence to support this conclusion knowing that their recommendations would fail to solve the problem but would avoid criticism from the manager.

Setting aside these human shortcomings, properly applied investigative techniques can be of considerable value.

ROOT CAUSE FAILURE ANALYSIS

Root cause failure analysis (RCFA) includes actions to determine both why a failure happened as well as the corrective actions to eliminate the failure or reduce its impact (Figure 10-18).

FAILURE MODES AND EFFECTS ANALYSIS

The failure modes and effects analysis (FMEA) studies failure causes and ranks their risk. Then the best technology is applied to reduce occurrences while improving detection capabilities. Failure mode is the way failures occur, and effects analysis relates to the consequences of the failures. Analysis classifies the severity of the failures on the equipment's system, such as a pump failure that shuts down an entire wastewater drainage system. FMEA anticipates what could go wrong by identifying the local effect (the immediate impact on the equipment that failed) as well as the next higher level effect (the impact on the equipment's system) and the end effect (the final result of the equipment failure). Each effect discovery leads to determining the best condition-monitoring technique to defeat the potential failure condition. A risk priority number (RPN) is assigned to the failures related to equipment being studied to establish the priority of corrective actions. This is the RPN computed as severity × occurrence × detection (Figure 10-19).

RISK-BASED INSPECTION

Risk-based inspection guides decisions in selecting equipment that poses the greatest risk if inspections are not done by ranking equipment according to its probability of failure and failure consequences. Then the most essential inspection services are contrasted against the ranking of equipment to allow the best allocation of resources to meet the basic inspection needs. The principal application of risk-based inspection is piping (Figure 10-20).

THE FINAL COST OF MAINTENANCE

Ultimately, the total cost of keeping production equipment in reliable condition is influenced by the cost of failures not averted. These create a multitude of costly consequences such as injury, lost product, and equipment damage. They represent 300% more than the cost of maintenance that could have avoided them (Figure 10-21). Most competent maintenance programs act to reduce the cost of maintenance. But the successful application of RCM attacks the cost of failure. When these failures are avoided or mitigated, the increased reliability of equipment represents a significant cost savings that is ultimately reflected in plant profitability.

SUMMARY

Reliability centered maintenance adds a valuable new dimension to the conduct of industrial maintenance. It focuses on the avoidance of failures and their far-reaching and costly consequences. Reduction of the massive cost of failure

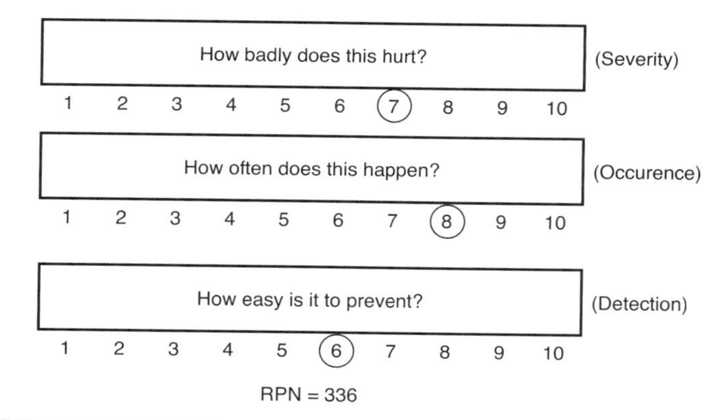

FIGURE 10-19 The product of severity, occurrence, and detection yields the risk priority number. Often, an RPN target is set after severity and occurrence ratings are established from historical equipment performance data. The challenge is to identify the best condition-monitoring techniques to meet the RPN target. This step requires a deliberate assessment of the detection capability of the total PM program.

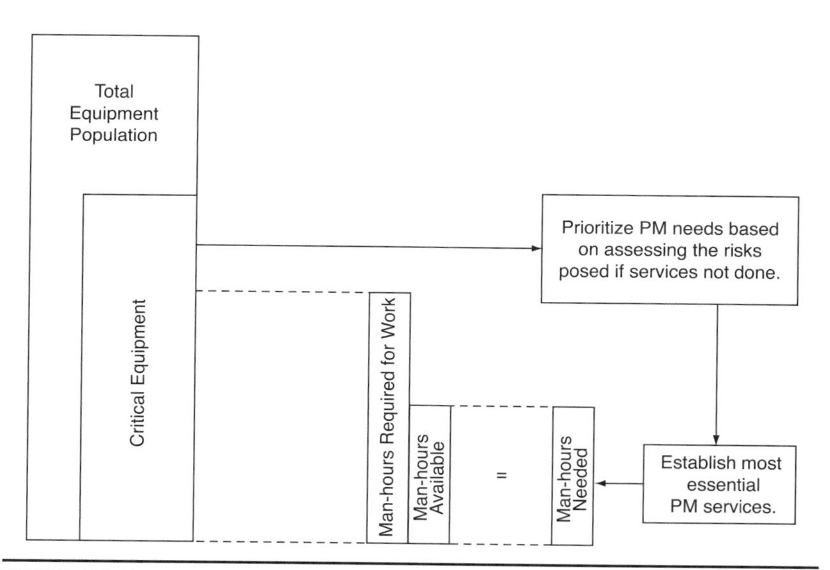

FIGURE 10-20 Risk-based inspection has the objective of determining the best allocation of resources for conducting PM services of piping applications. After critical equipment is identified, all applicable PM services are listed and their labor requirements established. Should labor needs exceed availability, the most effective PM services are identified and applied consistent with available labor.

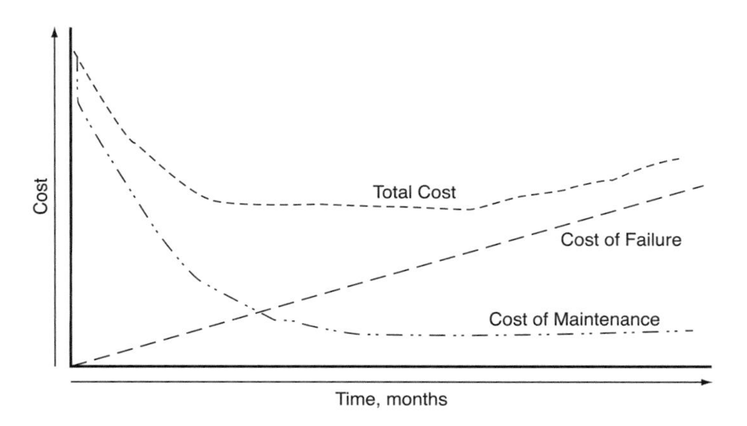

FIGURE 10-21 Total costs in relation to each other

becomes the most effective means of using a maintenance strategy to help achieve plant profitability. When coupled with equipment management that has brought all departments together in helping one another, RCM can add substantially to the achievement of equipment reliability.

REFERENCES

Allen, T. 2006. RCM, the Navy Way for Optimal Submarine Operations. RCM Forum in Las Vegas, NV.

Moubray, J.M. 1994. *Reliability-Centered Maintenance.* Oxford: Butterworth-Heinemann.

———. 1996. Redefining maintenance. *Maintenance Technology Magazine.* (April):21.

Nowlan, F.S., and Heap, H.F. 1978. *Reliability Centered Maintenance.* Washington, DC: Office of Assistant Secretary of Defense.

CHAPTER 11

Total Productive Maintenance

OBJECTIVES

Total productive maintenance (TPM) brings together the American practice of preventive maintenance (PM) with the Japanese concepts of total quality control and the involvement of all employees. It has the objectives of

- Maximizing equipment effectiveness,
- Establishing PM activities for the entire life-span of equipment,
- Involving all departments in a joint effort,
- Gaining the input of every employee from manager to worker, and
- Promoting itself through small-group activities like operator maintenance (Nakajima 1988).

TPM has its widest application in manufacturing. The most practical aspect of TPM for the mining industry is the application of autonomous maintenance, the direct participation of equipment operators in maintenance-related actions such as cleaning, adjusting, inspecting, calibration, and, under certain circumstances, minor component replacement. TPM tends to look at equipment operators as those who might apply "first aid" to ailing equipment while professional maintenance personnel are seen as the "surgeons"—performing the more complex maintenance activities such as overhauls or major component replacements. In the mining industry, the idea of operator maintenance moves between two wide extremes. Underground operations seem to have the widest application because mechanics are already part of operating crews. As such, it is not unusual to see mechanics or electricians operating equipment and operators assisting operators to get a unit of equipment back into operation. Open pit mining is another place where operators are of great help in making before-, during-, and after-operations checks of equipment. Plant operations tolerate operator maintenance, but maintenance personnel often prefer that operators stay in control rooms with maintenance coming to them if necessary. Of the various maintenance-related actions that operators could take, maintenance requires that actions be taken correctly and carefully.

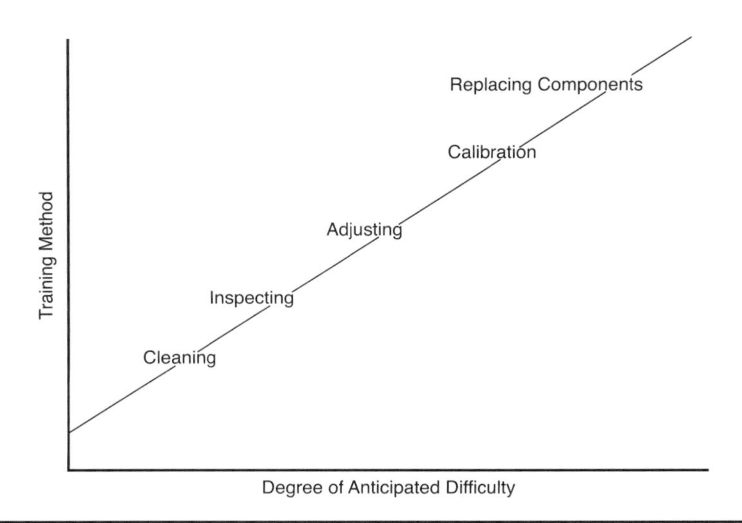

FIGURE 11-1 Stages of operator maintenance. If operator maintenance can help, maintenance must assess operator capability and training needed.

There is always some degree of maintenance concern that operator misunderstanding might create equipment problems rather than solve them, such as described in the following example:

> In one plant, the operating crew in the control room did a wonderful job repainting the floor with a tough epoxy finish. But maintenance appreciation turned to grief when they discovered that the control room crew stored the highly flammable paint cans behind sensitive, electrical control room panels.

After maintenance assesses the degree to which operators could help with maintenance, the degree of difficulty of anticipated tasks and the training needed or able to be absorbed by operators must be carefully considered (Figure 11-1).

TPM also brings an acute awareness of preventive maintenance to the total mining operation. In turn, key personnel can have a more direct impact on the quality of maintenance, as portrayed in the following examples:

> A mine manager was disappointed with only 14% compliance with the weekly PM schedule. At the next morning meeting she announced that any operating supervisor who could not make his equipment available to meet his 85% schedule compliance target would henceforth be required to explain to her personally why he could not. From that day forward, compliance was a genuine 85% and it remained there. Emergency repairs were quickly reduced as well.
>
> A tough vice president of operations ran a daily inquisition of his mine managers, publicly chastising them for their failures to meet

production targets in his dreaded conference calls. An interested consultant suggested that the vice president extend his inquiries to PM schedule compliance. Initially, none of the mine managers knew. But the next day they did, and with a new inquiry target, the vice president and the collective mine managers quickly realized that missed production targets and a poor PM schedule were closely related. A vastly happier vice president and a relieved group of mine managers had just learned why quality PM is so important.

TPM emphasizes the condition of a plant's production equipment. The condition influences how well worker productivity, cost control, correct product inventories, safety and health, productive output of equipment, and quality can be achieved. In turn, these factors impact profitability. And, in an age of more sophisticated equipment, automation will be used. Automation ensures that the tolerances, speeds, pressures, and temperatures required in the production process are met precisely. Thus, the expectation of unstaffed production could reduce the need for labor. However, the conduct of maintenance will still require labor, but the amount and intensity of its use can be minimized with better maintenance practices. Therefore, TPM places its emphasis on preventive maintenance to improve equipment effectiveness and tries to involve all employees in the effort. TPM also targets zero breakdowns and zero defects. As defects are reduced, the availability of equipment will increase while costs are reduced, product inventory is minimized, and productivity is increased. Although zero defects are sometimes possible in a confined manufacturing operation, they are less likely in the hard-use mining environment.

The implementation of TPM requires several years. During the initial phases, equipment must be restored to a safe, effective operating condition and personnel educated on proper maintenance practices. Then as the benefits of TPM are accrued, the initial costs of restoring equipment and educating personnel yield better productivity. In minerals processing operations, annual shutdowns attempt this periodic restoration process, but the practice is not as widespread as in other heavy industries such as oil and gas production. Finally, the initial investment costs are displaced with profits derived from better equipment condition and more efficient operation. TPM is often referred to as "profitable PM."

DEVELOPMENT OF TPM

The desire to escape from the excessive expense of the repair and downtime costs of breakdown maintenance constitutes the first developmental stage of TPM. In turn, the search to find a better way leads to stage two: implementing basic preventive maintenance, largely periodic servicing, and overhauls. Productive maintenance, or the combining of good initial equipment design (low maintenance), the application of techniques to improve reliability like condition monitoring, and the continued emphasis of solid maintenance practices constitute the third stage. But in the fourth stage, preventive maintenance and productive maintenance are brought together by the involvement of all employees.

TPM consolidates the lessons of preventive maintenance and productive maintenance. Then it commits the total plant population to continued support through education. Thus, TPM has become productive maintenance involving total participation (Nakajima 1988):

$$\text{TPM} = \text{preventive maintenance} + \text{productive maintenance} + \text{autonomous maintenance}$$

in which

- Preventive maintenance is the use of inspections, testing, and monitoring to avoid premature failure plus lubrication, cleaning, adjusting, calibration, and the replacement of minor components to extend equipment life, all applied throughout the life cycle of the equipment;
- Productive maintenance includes equipment design to minimize maintenance and emphasis on preventive maintenance, major component replacements, and overhauls combined with repair techniques and equipment modifications to preclude unnecessary breakdowns and make maintenance easier; and
- Autonomous maintenance is the *unique* aspect of TPM. It is the small group activities of cleaning, adjusting, calibration, lubrication, and minor repairs carried out by operators. It adds a dimension of motivation created when operators contribute more directly to maintenance. This point is the most significant one for mining since maintenance is already a difficult activity; therefore, extra help is appreciated.

TPM aims to maximize equipment effectiveness, establish a life-span PM system, involve all employees, and promote itself through the motivation afforded by small-group activities like operator maintenance. The use of the word total in "total productive maintenance" suggests three meanings:

1. Total economic efficiency in every phase of plant operation;
2. Total maintenance programs to ensure maintenance avoidance, better maintenance, and improved maintainability; and
3. Total in that everyone participates.

EQUIPMENT EFFECTIVENESS

Equipment effectiveness means production improvement to maximize output and minimize input. Output includes productivity, quality, cost, delivery of product, safety, health, environment, and morale. Input is money, labor, machinery, and materials.

Automation causes a shift from workers to machines, suggesting that the output is affected more by equipment condition than the efforts of workers. Therefore, TPM tries to achieve overall equipment effectiveness by minimizing the "six big losses":

1. Equipment failure
2. Lost time for setup and adjustments
3. Idle equipment and minor stoppages
4. Reduced equipment speed
5. Process defects
6. Reduced equipment yield

In determining equipment effectiveness, the six losses must also be applied to ensure that true equipment effectiveness is determined. If, for example, scheduled operating time is 100% of available time, true equipment effectiveness must also consider the reductions attributable to downtime, speed losses and defect losses. Collectively, these losses reduce equipment effectiveness further (Nakajima 1988). In mining, downtime rather than speed losses and defect losses are the biggest contributors to reduced equipment effectiveness. In turn, reduction in downtime is most influenced by better preventive maintenance. There are fewer instances of reductions in actions taken to reduce speed losses or defect losses in mining because most production equipment is put into operation exactly as the manufacturer designed it. Manufacturing industries, on the other hand, favor equipment modification as a means of improving speed and defect losses. But the most effective impact in mining maintenance comes from a competent program supported by the full support of the total work force as emphasized by equipment management. Thus, TPM makes its best contribution to mining with autonomous maintenance, increased awareness of the value of preventive maintenance, and by carrying out an equipment management–oriented maintenance program.

TPM AND EQUIPMENT FAILURES

The traditional failure curve shown in Figure 11–2 suggests early failures (A) followed by a period of random failures (C), ends in a wear-out failure period (B). TPM suggests that better preventive maintenance (PM) can reduce early failures while better maintenance can reduce failures during the wear-out period. Thus, blending RCM emphasis on the use of condition monitoring to control the incidence of random failures focuses the best of these two strategies on the problems of reducing failures and their consequences during the whole life cycle of mining production equipment.

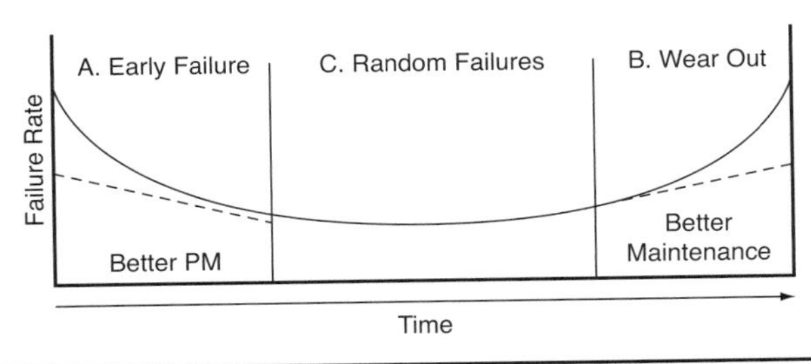

FIGURE 11-2 Even with the reduction in failures from better preventive maintenance during the early failure period and better maintenance during the wear-out period, TPM suggests that better equipment design and condition monitoring during the startup and random failure periods are necessary. In turn, the increased use of condition-monitoring techniques during the period of random failures strengthens the whole mining maintenance effort.

Therefore, TPM suggests "countermeasures" to minimize failures. Essentially, a significant number of breakdowns can be minimized through the maintenance-free design of equipment. But that condition is far from realistic despite the gains that are being made in building greater reliability into modern production equipment. The reality is that most equipment in operation today is not in perfect condition. Therefore, the first step of the countermeasures to minimize failures must be to eliminate failures of equipment currently in operation with solid maintenance practices. Then based on the experience gained, the lessons can be used to improve the design of equipment and gradually bring it closer to a lower maintenance requirement with a future "maintenance-free" design in mind.

Failures are associated with functions, as in reliability centered maintenance (RCM). Thus, the loss of a "standard function" would mean that the failure is not limited to unexpected breakdowns leading to complete stoppage (the functional failure of RCM). Therefore, even as the equipment is running, various forms of deterioration can still cause losses.

These losses are manifested in the six big losses of TPM, including longer and more frequent adjustments, frequent idling, and minor stoppages, as well as reduction in processing speed and cycle time. These losses must also be treated as failures. Again, they are similar to the approach of RCM.

TPM classifies unexpected breakdowns with complete stoppage as "function-loss failures," whereas those lesser failures due to deterioration are classified as "function-reduction failures." Both TPM and RCM have thought through the relationship of various types and degree of failures. In addition, similar careful consideration has been given to the countermeasures of TPM versus the condition-monitoring programs of RCM.

Fundamentally, both TPM and RCM have the common goal of improving equipment reliability and the economic, performance, and productivity gains that such status implies. However, RCM takes a slightly different approach than does TPM. After the types of failures are carefully identified and the consequences of their failures assessed and prioritized, a program to minimize failures is established. In that program, RCM advocates the application of condition monitoring to observe equipment performance and avert random failures. Equipment is left in service on condition that the monitoring devices can signal that the functions of the equipment are still being achieved. In so doing, RCM can ensure that major components, for example, which would normally be replaced at the end of a traditional wear-out period (as in TPM), can be "monitored" and left in operation longer. In turn, these gains are a source of longer component life, cost reduction, better productivity, and ultimately profitability. TPM, on the other hand, focuses on the identification of hidden defects and their correction before equipment breakdown.

TPM advocates five countermeasures to help reduce or eliminate failures (Nakajima 1988):

1. Keep equipment in basic good condition with cleaning, lubrication, adjusting, and other autonomous maintenance activities.
2. Operate the equipment properly.
3. Restore deterioration by performing effective maintenance.
4. As maintenance is being carried out, observe design changes that could be incorporated into future equipment and incorporate them toward the future objective of a maintenance-free condition.
5. Improve operation and maintenance skills through training and education.

Both TPM and RCM would advocate all of these measures, but RCM would be very precise on the exact means by which maintenance activities emphasizing condition monitoring would be carried out.

LIFE-CYCLE COSTING

The determination of life-cycle costing or the comparison of the cost of performing maintenance over the life-span of equipment is another important TPM consideration. It is used to determine how to do maintenance as well as which equipment may be less costly to operate and maintain.

A decision whether to overhaul a haulage truck is a good example of life-cycle costing. This decision is forced when the truck has so many things wrong with it that it must be removed from service. Among the factors are cost, downtime, number of repairs, frequency of repairs, seriousness of repairs, and reduced output of the truck. When a decision has been made, the truck is subjected to a pre-overhaul inspection and all major components are tested to determine whether they can be reused, rebuilt, or replaced with new components. Then the unit is reassembled, tested, and restored to service, as shown in Figure 11-3. Life-cycle costing compares the cost of the overhaul and its

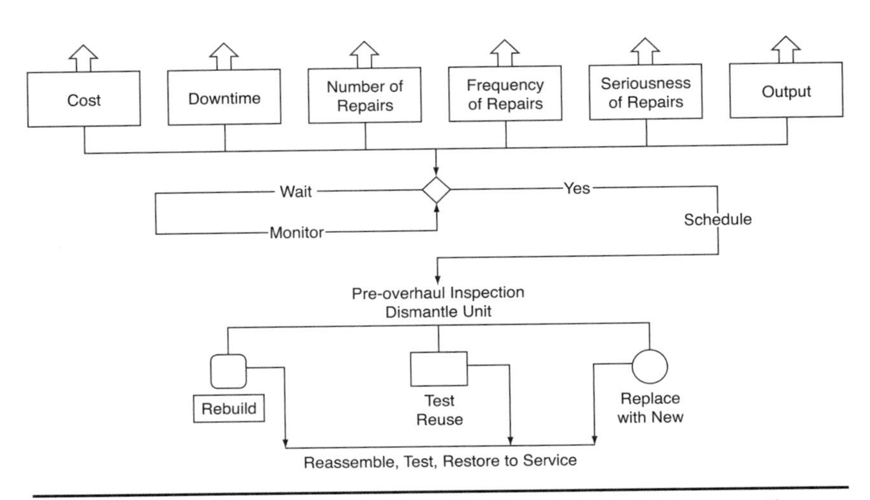

FIGURE 11-3 Overhaul decisions illustrate life-cycle costing by contrasting their cost-effectiveness with a major component replacement program

capability of restoring the truck to an as-new condition versus the continuous replacement of failed components on trucks not being subjected to overhauls. In most instances, the overhaul is more cost-effective because it deals with the total truck, whereas major component replacements focus on a single component while possibly overlooking more serious problems that the overhaul would uncover.

Life-cycle costing should be reflected in each of three ways:

1. Present value—Convert all earnings and expenses into current money values.
2. Final value—Sum up all expenses and investments at the end of the comparison period.
3. Annual value—Calculate the annual cost using the value of money year by year.

In each instance, the initial cost of equipment, the cost of financing its purchase, and the cost of maintenance are factors in life-cycle costing.

IMPLEMENTING TPM

Requirements for the successful implementation of TPM include

- Taking steps to reduce the six big losses,
- Initiating autonomous (operator) maintenance,
- Applying a competent scheduled maintenance program,

- Increasing the skills of operators and maintenance personnel, and
- Applying these factors to specific types of equipment (Nakajima 1988).

These steps are accomplished in a systematic way. For example, equipment failure reduction is initiated by restoring equipment. Then autonomous maintenance is started with basic cleaning, addressing sources of problems, and establishing standards for lubrication and cleaning. These actions tend to reduce the variability of equipment life-spans because equipment problems are being more carefully identified by operators and corrected more quickly. Next, a vigorous general inspection program for equipment is initiated using the PM program to control the actions. This step lengthens the life-span of equipment because it is the beginning of a competent, scheduled maintenance effort. This is now followed by the introduction of autonomous inspections by operators in which more personnel are watching the equipment. This results in improved day-to-day care of equipment by operators. The end point is that only occasional repairs are required as a result of the dual application of autonomous operator maintenance attention and a regular maintenance program.

Finally, with the introduction of autonomous (operator) maintenance, equipment life-spans become very predictable. The resulting orderliness and better organization of the total maintenance effort yields greater productivity, performance, and reliability.

Three requirements are necessary for improvement. First, the motivation of personnel must be increased to better enable them to reduce the six big losses. Second, their skills must be improved to raise their competency to maximize equipment effectiveness and operation. Finally, the work environment created by management support and enthusiasm must encourage individuals collectively to get the job done. Thus, the direct involvement of management is critical, as in the earlier example wherein the mine managers and vice president of operations discovered the real value of preventive maintenance. Twelve TPM implementation steps pass though three distinct stages:

- Preparatory stage—Establish a suitable environment by developing the plan for implementation, preparing personnel for the task, and establishing goals.
- Implementation stage—Establish the TPM program.
- Stabilization stage—Normalize the total operation by integrating TPM into plant life. It is now the best, most acceptable way to achieve profitability (Nakajima 1988).

Steps 1–5 constitute the preparatory stage of the TPM development program. Steps 6–11 constitute the implementation stage, and step 12 is the stabilization stage (Nakajima 1988).

Preparatory stage:

1. Introduce TPM along with concepts, goals, and expected benefits of implementation. Follow up by establishing a favorable environment for implementation.
2. Educate personnel to raise their morale and reduce resistance to adopting TPM. This step provides assurances that TPM will be beneficial and reduces resistance to change.
3. Create internal organizations to discuss problems that must be solved, suggest solutions, and talk them over from one organizational level to the other. This establishes the participatory management style of TPM. Groups will raise questions that sharpen the awareness of all personnel and cause others to suggest more effective implementation solutions.
4. Establish policies and goals that prescribe standards for interactions between groups. Based on the policies, departments can develop day-to-day procedures with specific, realistic, attainable goals in mind.
5. Formulate the master plan to establish a timetable for implementation actions that structures actions against goals. Use the plan to gauge progress, plan the next steps, and organize resources.

Implementation stage:

6. Formally launch the campaign, allowing management to report on the outcome of the preparatory stage, reiterate policies and goals, and expand on the details of the master plan. Provide opportunity for workers to make a firm, public commitment to successful implementation.
7. Improve equipment effectiveness with a collaborative effort of maintenance, operations, and engineering to find better ways to carry out maintenance. This step demonstrates how TPM has helped. Maintenance launches autonomous maintenance in a total commitment while efforts to measure results against goals demonstrate commitment of plant personnel.
8. Develop the autonomous maintenance program to demonstrate departure from the "I operate—you fix" mentality. Extraordinary steps ensure against regression to former behavior. They include
   a. Thorough operator cleaning to become more familiar with machines, provide an opportunity for closer examination, and create interest in the first level of equipment care—keep it clean!
   b. Establish procedures and methods to keep units clean.
   c. Establish standards of cleanliness and lubrication.

d. Conduct general inspections as part of the maintenance department's PM program. Use these inspections to determine the condition of equipment and establish how well the individual operators are performing tasks to date.

e. Replace the general inspections by maintenance with autonomous maintenance inspections now that operators know better what to look for.

f. Concentrate on organization and tidiness of the overall work place to create greater awareness of operating and working conditions. This step stimulates operators to make work more efficient.

g. Finally, implement autonomous maintenance fully. Make it a formal activity wherein checklists are used by operators, reports similar to PM compliance reports are made, and the work order system is used by operators to request help and keep a record of jobs they can do. Operators are now a part of the maintenance effort.

9. Develop and implement a competent, scheduled maintenance program.
10. Conduct training to improve skills. Specialized maintenance skills training is provided to maintainers to ensure they are competent in the latest, best diagnostic and repair techniques. Similarly, operators are trained in proper operating techniques and essential basic maintenance requirements.
11. Develop programs for specific types of equipment as maintenance and engineering combine to minimize startup problems.

Stabilization stage:

12. Perfect TPM implementation and raise TPM achievement levels. New personnel will join the organization, equipment will be replaced, and new processes introduced. And, with future implementation actions, more impressive goals will be established.

SUMMARY

Total productive maintenance reveals that the strength of any maintenance strategy is to choose complementing elements of approaches like TPM or RCM and apply them within the larger framework of equipment management.

REFERENCE

Nakajima, S. 1988. *Introduction to Total Productive Maintenance*. Cambridge, MA: Productivity Press.

CHAPTER 12

Implementing Information Systems

QUALITY TRAINING MEANS SYSTEM ACCEPTANCE

Acceptance of any new information system will depend on how well the users understand how it can help them to perform their jobs. Education of system users is a key factor in successful implementation. The maintenance manager should review the system training program to ensure it is complete. When training begins, the manager should take an active role and insist that each supervisor do likewise. This will demonstrate commitment to the system and encourage users to master its use quickly. At the outset of the training program, the duties of key personnel should be associated with the information they will need. Then those aspects of information that personnel must have to perform their duties should be emphasized in the training program. Maintenance supervisors, planners, and craftsmen function as a team, and because they will be the principal system users, they should be trained together. Because the supervisor is responsible for getting the work done on time and with a quality effort, work control is his or her primary responsibility. Therefore, the system training should explain how supervisors should

- Review outstanding jobs,
- Determine routine (dynamic) preventive maintenance (PM) services to be done,
- Make decisions on jobs to be done,
- Assign jobs to crew members,
- Provide job instructions,
- Monitor job status, and
- Close out completed, unplanned jobs.

Typically, the planners' training should include

- Use of work orders for planned jobs,
- Determination of work order status,

- Determination of work order cost,
- Observation of the backlog, and
- Observation of the use of labor.

To ensure that the work control task remains in the hands of the supervisor, the planner should focus on major planned and scheduled jobs. After a planned job has been placed on an approved schedule, the supervisor would make the work assignments for a specific day and shift and then assign it to crew members.

Typically, a PM service that requires no equipment shutdown can be accomplished anytime within the week it is scheduled. An individual crew member's training should include

- Entry and use of equipment specifications data,
- Initial reporting of labor data (verified by supervisor),
- Updating of equipment specifications data,
- Review of repair history data, and
- Entering of new work.

Training should be started with useful systems and information. Craftsmen need to know part numbers. Therefore, they should first be trained on using equipment specifications. Soon they will be using the computer regularly and confidently.

THE PROBLEM OF COMPUTER LITERACY

The degree of computer literacy will differ among personnel, which must be considered in training. Often, the planner will have had previous computer experience and possess administrative skills. Usually, electricians and instrumentation personnel will have been exposed to the use of the computer as a diagnostic tool and they can easily adapt. Mechanical craftsmen usually have less need for the computer in their work, but as younger craftsmen enter the work force, it is more common to see computer literacy based on their use of personal computers.

Some maintenance supervisors present a problem during system implementation because they are apprehensive about using a computer. Typically, when paper systems existed, the supervisor was unquestionably in charge. However, when the computer displaced these systems, many supervisors shared or yielded work control to planners. The supervisor's primary responsibility is work control. Training must give this aspect strong emphasis. Maintenance supervisors prefer an active field role over administrative matters. Yet, making job assignments for crew members and following up on job progress are essential supervisory tasks even though administrative in nature. Therefore, the supervisor cannot ignore the value of the computer. The best course of action is to install the system for the supervisor and crew as a team. Involve the craftsmen in entering new work in the computer, especially equipment deficiencies they have found during PM inspections. This way, expertise among crew members contributes to the supervisor's progress in effectively learning to use the computer. As the supervisor and crew start to use the computer, they will demonstrate its value to themselves and soon overcome reluctance to use it.

REPORTING FIELD DATA

Training should include the reporting of field data by crew members and its verification by supervisors. If the field data are incorrect or incomplete, so is the information produced. No matter how well-conceived an information system may be, the critical aspect of its success is the accuracy, completeness, and timeliness of the field data. Field data includes all information reported by all plant personnel as they carry out work or perform maintenance-related functions:

- Maintenance craftsmen report man-hours, record equipment numbers, describe repairs, indicate failure codes, etc.
- Maintenance planners open and close work orders.
- Warehousemen post equipment and work order numbers against stock issues.
- Purchasing agents transpose work order or equipment numbers onto purchase orders.

Supervisors verify most data, whether it is reported on a time card or on the work order format displayed on the computer screen. Errors can seriously affect the quality of information. The most effective way to ensure that field data are reported accurately and completely is to train the personnel who report it. The key person is the maintenance craftsman. He is the actual person who does the work and is in the best position to accurately describe what he has done. In most instances, the craftsman is willing to report data. However, he must understand the information that his data will produce, what it will be used for, and why it is important. Training for craftsmen must not be limited to showing them how to fill in the blanks on the format displayed on the screen. They must be shown what their data produces. Showing the craftsmen the actual reports and letting them use the display screens will contribute substantially to their appreciation of the importance of the information. Some maintenance managers object to crew members using the computer. They see it as a distraction from the time they "should be turning wrenches." This outmoded view overlooks the fact that crew members have the facts on work they have performed and will report accurately. Too often, data is distorted as it passes through supervisors, planners, and clerks on its way into the system database. The result is usually poor information and poor decisions. Moreover, with the advent of a younger, computer-oriented work force, many craftsmen are already skilled in the use of the computer; therefore, by denying its use, a vital source of quality information may be overlooked.

In a large work force, craftsman usually cannot be given individual training. A training procedure must be worked out. Often the best way to do this is to show the craftsmen a facsimile of the reports that will be produced and encourage some discussion on how the information can be used. As actual reporting begins, it is necessary to familiarize personnel with codes that will be used in reporting. Helpful techniques include

- Listing failure codes on the reverse side of a field reporting document, and
- Providing for selection from an on-screen table file.

These techniques help avoid unnecessary mistakes. It is also helpful in successfully initiating field reporting to accomplish it in phases. For example,

- If reporting will include components, it is best to omit component reporting until equipment reporting is mastered.
- If system development will be delayed, start practice reporting early to gain proficiency.

Training for warehouseman and other administrative personnel should not be overlooked. The main point to remember is that those who do not understand the purpose of what they are reporting will, most likely, not do it well.

THE SUPERVISOR AS TRAINER

One of the best methods of ensuring that craftsmen are well-trained is to require supervisors to carry out the training. Supervisors are motivated to do a good job because they know that crews will be a critical audience. The best way to find out whether supervisors understand the system is to listen to them explain it to their crews. As they do so, the quality of their understanding as well as their commitment can be observed.

USING THE PLANNER

Maintenance planners will often have been involved in system development and be familiar with details. They can be used to help explain data entry requirements to supervisors and material control personnel.

DOCUMENTATION

Training materials must be well-documented and easily available to personnel. Table files from which the user can make selections are very helpful, as are help screens which offer clarifying notes on system use.

LOADING FILES AND SYSTEM TESTING

All of the files should be loaded before training begins. There is nothing as confusing as the attempted use of "bogus" data, equipment numbers, or descriptions for the person learning to use a new system. Use the actual equipment numbers and work descriptions. Training is more realistic when actual data is used. Prior to loading files in the programs, test data should be tried to ensure that the data is properly transported between systems and that it produces the desired information. System files should be loaded carefully, especially equipment numbers and components. Smaller existing files such as failures codes usually present no problems. Assistance may be needed from the information department when previous files such as repair history or equipment specifications (part numbers, drawing references, etc.) must be converted and transferred.

VENDOR TRAINING

Many vendors of software packages may not have had specific experience in industry. The worst vendors will explain the system to the clerk and leave. The best ones will load the files and train everyone because they know they have a quality product and they want a solid endorsement from a satisfied customer.

PRODUCTION PERSONNEL

Production supervisors will be required to enter new jobs into the system and make inquiries on the status of jobs. They must be included in the training so they can do these things competently. Production managers will require information on performance indices and costs. They should be shown the reports and have the use of the information explained to them. By starting as early as possible on the training of production personnel, they will be in a better position to offer comments on how the information can help them. Often, these comments will influence alterations in the basic maintenance program as it is modified to utilize a higher quality of information.

DEVELOPING A PLAN FOR INFORMATION USE

The vital questions of who gets what information, when, or how often they receive it, as well as what actions they are expected to take must be addressed. This makes the training practical and realistic. The following information uses are typical:

- The maintenance manager requires summaries that describe overall performance and cost control, allowing him or her to judge accomplishments and direct corrective actions.
- General supervisors need details of weekly accomplishments like PM schedule compliance to direct improvement actions.

- Supervisors need work details to develop jobs for their crews along with feedback on job progress.
- Planners need to know the status of major jobs and to explain variances from their recommended job plans.
- Craft personnel need to be able to look up part numbers and query repair histories to prepare for upcoming jobs.
- Maintenance engineers must track costs to find troublesome equipment, narrow problems with repair history, then field-check to develop corrective actions.
- Operations managers want to know how much maintenance services have cost and obtain an explanation of current and future costs.
- Operations supervisors will want to enter new jobs into the system and determine the status of outstanding jobs.

SUCCESSFUL IMPLEMENTATION STEPS

A competent maintenance information system could fail to deliver quality information if its users are not properly prepared to utilize its capabilities. As a result, information is incomplete, of poor quality, and decision makers could fail to control work properly. The following ten steps can help ensure this does not happen.

Step 1. Confirm the soundness of the equipment management program. Most unsuccessful attempts to implement information systems can be traced to inadequate program definition and a lack of education of the personnel who must carry out the program. Even if the most effective information systems are superimposed on a confusing or ill-defined program, they will fail. The information system is merely the communications network of the maintenance program. Therefore, verify that the maintenance program is properly defined before commencing system implementation.

Step 2. Ensure that roles of key personnel are correct and understood. Information systems have three objectives: identify work that is required, control the work, and measure the success of the work performed. As the maintenance program is executed, key personnel perform one or more of these functions. Thus, the information system provides the tools by which they carry out their roles. All users of the work order system define work through the work order system. Some work is identified by the superintendent as she observes high costs to identify equipment requiring attention. The maintenance engineer identifies other work based on failure trends summarized in repair history.

Work is controlled by the maintenance supervisor as he directs the efforts of his crew. The success of work performed is measured by a planner who notes the comparative cost and timely completion of major planned jobs. Similarly, the superintendent notes cost variances while the supervisor observes the cost of units and components in his area of responsibility. A plant manager checks performance using indices like cost per unit of product. Specific information is directed to designated personnel who, in turn, make decisions and take corrective actions. Information needed by key personnel should relate to the actions required of them. Responsibilities for action must be confirmed before information requiring action is directed to these individuals. Thus, early confirmation of the roles played in managing maintenance will help in the effective utilization of the information.

Step 3. Phase out all previous, conflicting procedures to avoid unnecessary difficulties as you install any new information system. The new information system is now the "official" communications network of the maintenance program. If it has been carefully developed, it need not be supplemented by word-processed listings of jobs or PM services scheduled on a spreadsheet. Make a clean break and give the new system a chance to meet performance criteria unencumbered by outdated, ineffective habits and procedures. Be especially aware that personnel problems will require attention. The following situation is typical:

> One maintenance superintendent was surprised that a new, highly acclaimed information system failed to be able to schedule lubrication services. A closer investigation revealed that six competing procedures, archaic remnants from several previous programs, were still in use. Their continuing use precluded reliance on the new system and, as a result, services were being scheduled haphazardly with no followup on schedule compliance.

Any new, competent system must be given a fair test. To allow personnel to continue to use and depend on previous procedures denies that test. As a result, the new system may fail to deliver quality information.

Step 4. Load all files and modules, and confirm the equipment numbering scheme. Training will be confusing if personnel are asked to pretend that, for instance, "ABC" is a conveyor. Therefore, load all files and modules before training begins.

Unless the training yields immediate, useful results, the new program will be off to a shaky start. Make certain that the equipment numbering scheme is sensible. To illustrate,

> Two processing plants built 6 years apart used the construction codes of two different contractors to number their equipment. Not only were the numbering schemes different, but each failed to divide equipment by type, unit, and component. A subsequent attempt to install a new system was delayed considerably while steps were taken to realign equipment numbering procedures into a common arrangement.

The focal point of all maintenance work is equipment. If the work on equipment is to be effectively managed, equipment must be given a logical numbering scheme. This will make the implementation of any information system much easier.

Step 5. Verify that the hardware and networking arrangements are functional. If the system won't convey information adequately, not only does this situation create doubt in the minds of its users, but it will set the implementation timetable back considerably. This is avoidable. Test and then test again. Make sure that everything works properly.

Step 6. Verify that the field data which will fuel the system are complete, accurate, and timely. The work order system is the means by which field data are directed into the information system and converted into useful decision-making information. However, if these data are incomplete or incorrect, little worthwhile information will be created. In addition, still fewer correct decisions will be made or proper actions taken. Quality field data is the catalyst of good information. To illustrate,

> In a large plant complex, maintenance supervisors "created" and entered the labor data for their crews based primarily on assignments made at the start of the shift. This procedure neglected all of the verbal orders crew members received as well as the work they identified and completed on their own. Review of the resulting cost reports often showed massive amounts of material installed "without the aid of human beings"—material was charged but no labor was reported. The supervisor's labor data was conjecture and misleading. They understood how they could manipulate the system to simplify labor reporting but, in the process, revealed total ignorance about the value of quality information.

In this instance, crew members, as the best source of information on work they performed, should initially report the field labor data. The supervisor should verify it. Don't waste time, money, and energy on the installation of a new information system unless you are able to develop the best field data you can get.

Step 7. Establish a core group of maintenance personnel that will train the work force. The "core" group must receive an in-depth education on the entire information system because they must be able to answer any question on the total system. Also, prepare them for unusual situations, because they may encounter resistance and criticism that, while not aimed at them, may affect their objectivity in getting on with the training. Guidance on training by a system vendor can be useful, but training by company personnel will ensure it is practical and realistic.

Step 8. Conduct adequate training. The amount of training varies according to the backgrounds of individuals. If you are training an instrumentation crew that uses the computer daily in their work, expect to spend a week with them. If neither the supervisor nor the crew has ever used a computer, expect to spend several weeks with them. These times assume that personnel have a well-defined maintenance program and they understand it. It also presumes that the information system is easy to use. The core (in-house) team should move from area to area. Training should include 8 hours of classroom familiarization education per individual, followed by at least 21 consecutive days, around the clock, working with the personnel in the area. If you do not supply this minimum training, then you must be prepared to accept mediocre system performance as a result of "trial and error" substituted for quality training.

Step 9. Develop an implementation schedule with specific performance-related objectives. Set reasonable targets with realistic objectives. The following goals are typical:

- At the end of the first week, every crew member will be able to enter new work in the system, print work orders assigned to them, and report actual hours against each work order.
- At the end of the first week, each supervisor will know how to prepare a crew schedule assigning each member a full shift of clearly identified jobs.
- On the completion of the second week of training, each supervisor will be able to query repair history and review costs against jobs, equipment, and cost centers.

Don't omit anyone from the training. Since the information system is the "official" communications system, all employees must be able to use it competently. Consider the following scenario:

> Nine weeks after system implementation training was completed, a maintenance general supervisor was asked to pull together the repair history for a troublesome unit. When his information was questioned, it was discovered that he obtained most of it from handwritten log books rather than from the more complete, accurate, and up-to-date repair history of the new system. Further questioning revealed that he knew little about the system. He had excused himself from the training, expecting that others would produce the information for him when he needed it.

Unfortunately, some of his supervisors also assumed a casual attitude based on their observation of his actions, causing still further problems. Train everyone; no exceptions.

Step 10. Monitor system use and accomplishments. System performance will be important. It is not enough to create information. The information must be put in the hands of the proper people who, in turn, must make decisions and take corrective action. Monitor what happens. For example,

- If compliance with the PM program is improving, then confirm that fewer emergencies are occurring and more work is being planned and scheduled.
- If more work is being planned, see if worker productivity is improving.

- If productivity is supposed to be better, find out whether fewer overtime hours are being used.
- If compliance with the weekly schedule is not improving, then be able to pinpoint the cause:
  - Maintenance didn't get materials.
  - Production didn't release units.
  - Maintenance failed to allocate labor for jobs.
- If selected units show greater maintenance costs, expect to find meaningful narratives of what happened in the repair history.

WORK ORDER SYSTEMS

The work order system is part of the information system. In combination with the information system, work orders are the means by which maintenance carries out the steps that make up their program: requesting, classifying, planning, scheduling, assigning, controlling, and measuring work. Work orders constitute a system in which a specific type of work order is used to control each type of work performed. A formal work order, for example, is used to control complex planned and scheduled jobs that often utilize several different crafts, may require materials from several different sources, and where the cost and performance must be measured. By contrast, a standing work order is used for routine, repetitive actions like grass cutting or shop cleanup. Verbal orders might be used to initiate emergency repairs. The work order system also links with elements of the accounting system so that labor or material data can be obtained and matched with the work identified by work orders. When the work order is opened, it becomes an accounting document, and the combined data are then processed by the programming details of the information system. The result is decision-making information used to manage maintenance.

The work order system is not exclusive to maintenance. It also covers work performed by contractors and is also used by operations personnel to identify work they need and tasks they must perform themselves. The work order system should

- Cover all types of work performed, including engineering projects;
- Be able to document the job cost and performance of individual jobs;
- Allow the accumulation of costs to jobs, equipment, or functions;
- Provide for the control of routine, repetitive functions;
- Allow for the use of verbal orders (with followup);
- Provide suitably simple controls for simple actions like shop cleanup;
- Provide controls that are suitable to the kind of work being done;

- Link with production statistics to establish performance information; and
- Link with accounting documents to produce resource use information.

Work orders allow maintenance to

- Order service;
- Identify work required;
- Relate work to a unit of equipment;
- Identify functions to be performed;
- Establish the type of work, its timing, and importance;
- Identify labor, material, and equipment requirements for jobs;
- Make provisions for job approval;
- Tie field data to accounting;
- Facilitate the use of standards;
- Provide data for the accounting system; and
- Serve as the basis for scheduling work.

WORK ORDER ELEMENTS VERSUS TYPE OF WORK

Each element of a work order system has a practical relationship with the type of work it controls and the degree of control required. Jobs of short duration, for example, need little labor, are inexpensive, and require minimum control. Thus, a simple work order element is dictated. If the work order element for controlling simple work is too complex or a work order is required for every trivial job, verbal orders may predominate and valuable information can be lost. There should be separate elements of the work order system, each intended to control a certain type of work, but in the simplest, most direct way. Work order systems must include elements to cover all types of maintenance work as well as nonmaintenance work (e.g., construction).

EXAMINING THE WORK ORDER SYSTEM ELEMENTS

Each work order system element should have a specific work control purpose:

- Maintenance work order (MWO)—Controls complex, detailed, planned, and scheduled work.
- Maintenance work request (MWR)—Requests and controls simple unscheduled or running repairs.
- Verbal orders—Used to initiate emergency repairs, allowing reaction in the shortest possible time. Followed up with a work request or work order when there is a need to determine job cost and repair history.

- Standing work order (SWO)—A reference number identifying routine, repetitive actions such as PM services or shop cleanup.
- Engineering work order (EWO)—Used to control engineering project (nonmaintenance) work like construction that might be done by either maintenance or a contractor, or both working together.

Maintenance Work Order

The MWO is used to plan and schedule major jobs such as overhauls or replacement of components like drive motors. The MWO establishes estimated resource needs, job cost, and duration. As work is performed, labor and material data are collected to provide job cost and performance (Figure 12-1). When the job is completed, the MWO is closed, and cost and performance are compared with estimates, then summarized. Next, the job cost is closed against the equipment on which work was performed.

```
                        MAINTENANCE WORK ORDER
MWO#      Sec      Unit      Comp      CT        Unit Name     Comp Name
426820    024      CM01      100       S         Cont. Miner   Cutting

Description of Work: RPL CUTTER HEAD MOTOR +

St     Opn    Stt    Cls    Pr     Wa     FC     RH     SN/In    SN/Out
S      0610   0612   0612   90     No     8      X      128875   122889

Work Order Notes                                          EST       ACT
                                  Mechanics                12        12
See Material List                 Electricians
                                  LABOR                   300       300
                                  MATERIAL              25600     25720
                                  TOTAL                 25900     26020

Requested:               Approved:                Completed:
B. Easton                J. Southworth            T. Washburn
04/03/09                 04/16/09                 04/18/09
```

FIGURE 12-1 Sample maintenance work order

Maintenance Work Request

The MWR is used to control unscheduled repairs (running repairs). These are simpler, less costly jobs on which it is not necessary to isolate cost and performance. However, the jobs must be accounted for and many will contain information that should be placed in the repair history. Simplicity and ease of use are important.

Verbal Orders

Verbal orders are used primarily when the urgency of work does not permit time to submit a work request or prepare a work order. Verbal orders are a fact of life for most plant maintenance departments. “Outlawing” verbal orders seldom works. Provision should be made to ensure the work is effectively controlled and information about the job is not lost. When verbal orders are used, there should be a means for the maintenance worker to report what he or she has worked on. If it is subsequently determined that the cost of the job is required, a work order can be opened after the work is done but before the labor data is processed. Generally, a specific procedure should be provided to guide personnel in the use of verbal orders.

Standing Work Order

The SWO is used to control routine, repetitive functions like training or shop cleanup, or to identify groupings of equipment that exhibit a low maintenance cost, such as replacing light bulbs in offices. SWOs use reference numbers to charge an activity to the department in which it took place. In the following example, 9 signifies that the work order is an SWO. Then 01 indicates that the activity took place in department 01. Finally, numbers like 991 indicate the action taken (roof repair). The numbers should not be used to describe a major unit of production equipment, such as a haulage truck, as this would then require an SWO for each component like engines or tires.

SWO Number	Description
901986	Training
901987	Shop cleanup
901990	Building repair
902991	Roof repair

Engineering Work Order

The EWO covers capital-funded project work such as construction, equipment installation, modification, or equipment relocation. The EWO is initiated by plant or mine engineering because it covers projects initiated by them. Maintenance links its MWOs to the EWO while a contractor cross-references his purchase order (PO) to the EWO that controls the project (Figure 12-2).

VOLUME OF JOBS VERSUS COST

The greatest volume of jobs is done in less than 2 hours, yet this accounts for the least amount of maintenance cost. Such jobs should be controlled in the simplest way using, as appropriate, message forms or verbal orders. But big planned and scheduled jobs must be managed because they represent the major cost of performing maintenance work. Sixty-five percent of the volume of jobs performed by a typical maintenance organization is carried out in less than 2 hours. But these jobs account for only 35% of costs. About 35% of the total volume of jobs done is planned. It represents 65% of maintenance costs, which explains why jobs should be planned.

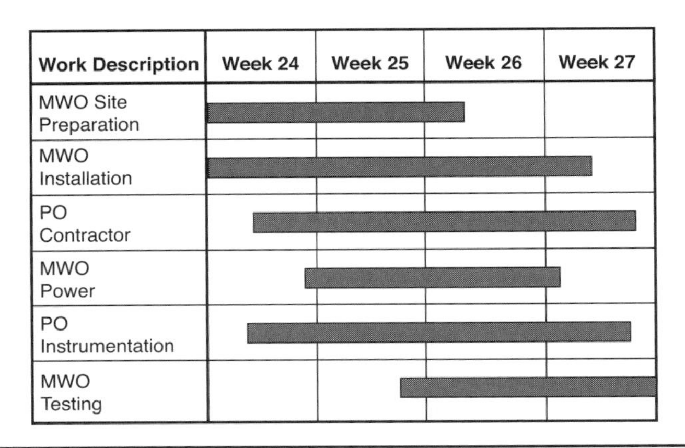

FIGURE 12-2 In this example of an engineering project being controlled by an EWO, maintenance and a contractor are collaborating. The contractor uses the PO for instrumentation work, while maintenance uses the MWO to perform site preparation or testing. Both the MWO and the PO are linked to the EWO so that the total cost and performance of the project can be managed.

SYSTEM INTERFACE

It is useful to observe the relationship between the elements of the work order system, the accounting system, and production control statistics as they work in concert to develop information. Data from these three systems are brought together, processed, and converted into information. In turn, the information is used to make decisions affecting the control of work, the use of resources, cost, and performance within maintenance and across the plant. When a work order number is assigned to a job, it establishes a temporary relationship with the equipment on which the work will be performed. It also establishes a relationship with details of the job, such as estimated hours by craft or estimated material cost. The work order permits the cost and performance of a job to be isolated. After materials are assembled, the work is scheduled and assignments are made to individual crew members to accomplish the work. Then as work is completed, man-hours are charged to the job by individual personnel. Thus, labor costs and man-hours accumulate against the job. It is then possible to compare labor cost and actual man-hours with estimates when the job is completed. After the job cost is compared with the estimate, it is closed against the equipment on which the work was performed.

EQUIPMENT NUMBERING

A logical and consistent numbering scheme should be applied to fixed and mobile equipment as well as to functions. The numbering scheme should include

- A department or cost center;
- A type of equipment or a function, like PM services, within the cost-center;
- Unit numbers for each piece of equipment;
- Equipment components as subelements of units; and
- Subdivisions of functions (e.g., PM Inspection, Route 3).

All of the equipment and function numbers are then brought together in a master equipment–function file, subdivided by cost center. Accounting must always be consulted to ensure that their asset numbering scheme is consistent with the equipment numbering scheme most useful to maintenance. Having numbers affixed directly on equipment should be avoided until the numbering scheme has been tested. Instead, a plant layout drawing should be used to provide a reference of equipment numbers while testing is in progress. A numbering scheme should show the cost center (CC), type of equipment (TP), unit number of equipment (UNIT), and components of the unit (CP):

```
CC   TP   UNIT  CP
06   HT   021   200  Haulage truck 21, drive line (200) in Cost
                     Center 06
```

FAILURE CODING

The failure code is used to identify the cause of failure for individual jobs. A typical failure code listing is shown in Figure 12-3. Over an extended period, the failure trend is analyzed for numerous jobs to determine a pattern of failures. The pattern, in turn, dictates the specific corrective action required.

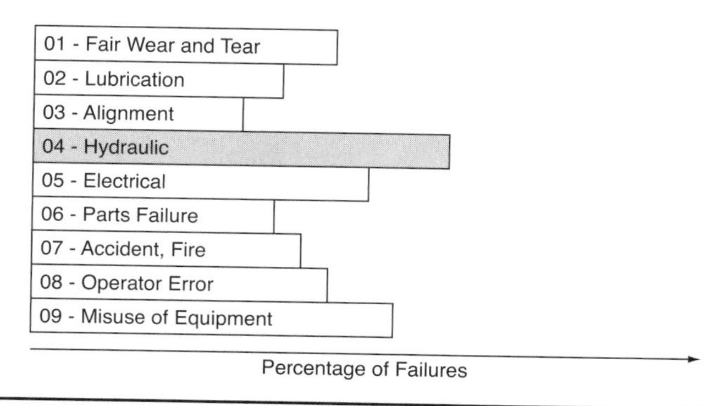

FIGURE 12-3 Hydraulic failures indicated the highest percentage of failures when a fleet repair history failure trend was analyzed in this example

WORK DESCRIPTION

Each description of work should be given a short title consisting of an action verb and an object. Other more elaborate descriptions may be added elsewhere in the work order as required. The short title provides an additional file search, allowing the development of detail for standardizing jobs. For example, of numerous possible jobs done on an electric motor, a standard job description of RPL ARMATURE (replace armature) would allow the isolation of all instances of this particular job. By isolating all work orders with this title, the computer can produce a standard bill of materials for this job. Also, a list of all man-hours by craft for every repetition of this job could be developed. As a result, labor standards for jobs can be constructed quickly. Typical action verbs and abbreviations are shown in Table 12-1. The use of standard titles also speeds the visual search of repair history because of the standardized verb–object combination.

TABLE 12-1 Typical action verbs and abbreviations

Verb	Abbreviation
Adjust	ADJ
Align	ALN
Clean	CLN
Fabricate	FAB
Inspect	ISP
Install	IST
Lubricate	LUB
Machine	MAC
Modify	MOD
Overhaul	OVH
PM inspection	PMI
Rebuild	RBL
Relocate	RLC
Remove	RMV
Repair	RPR
Replace	RPL
Reposition	RPN
Service	SVC
Splice	SPL
Test	TST
Weld	WLD
Winterize	WNT

TABLE 12-2 Priority rating

Production Rating	Maintenance Rating
10 One-of-a-kind equipment	9 To ensure safe operation
8 Reduce production capacity	7 Overhaul
6 Reduce output	5 Replace major component
4 No effect on production	3 Rebuild component
2 Improve working conditions	1 Paint, clean up
Priority number = Product of ratings	

PRIORITY-SETTING

There are two objectives of priority-setting:

1. Identify the importance of a job compared with all others.
2. Establish the time period within which the job should be completed.

By using the work category definitions, each type of work can help to define its own time frame within the priority-setting scheme. For example,

- An emergency repair would be completed within 1 day.
- An unscheduled repair would be completed within 1 week.
- PM services would be completed the same week as scheduled.

Using these categories of work and corresponding time periods, all work except scheduled maintenance can be provided a priority in the simplest, most appropriate way. Each planned and scheduled job would then be given a priority rating based on the importance of the equipment and the importance of the planned job being done (Table 12-2).

PRIORITIES APPLIED TO NONMAINTENANCE WORK

Nonmaintenance project work competes for the same craft labor used on major maintenance jobs. Therefore, there should be a comparable priority-setting scheme by which project work can be prioritized.

One of the major problems created with a poor priority-setting scheme is the buildup of work orders in the backlog. In some instances, jobs may have been in the log for many months. Generally, if a job has been in the backlog more than 16 weeks (4 months), there is about a 90% chance that

- The job has been accomplished piecemeal,
- It is no longer required, or
- The priority has changed.

Therefore, these jobs should be identified and reviewed with operations. Also, the backlog of all jobs should be purged after the first year of operation with any new system. This precludes poor estimates made while estimating techniques were still being learned from distorting future backlog information.

ASSEMBLING SCHEDULES

Schedules are made up of groupings of selected work orders. The focal point is usually a weekly schedule. Typically after a weekly scheduling meeting, all approved major planned jobs (and PM services requiring equipment shutdown) are assembled. Then labor is allocated, material delivery verified, and supporting equipment or transportation confirmed. The schedule is then given to the maintenance supervisors. In turn, they will add outstanding unscheduled jobs that can be done on the same equipment being shut down to perform major jobs or PM services. Thus, the collection of work requests that are ready can be added piggyback to the jobs on the approved weekly schedule. Using the weekly schedule as a guide, individual supervisors can prepare daily and shift crew assignments for their personnel. A daily coordination meeting with operations may occur to adjust the schedule in the event of delays. Using the information system, the status of jobs is observed along with the degree of job completion, accumulated costs on jobs, and so forth. Toward the end of the week, when the scheduling meeting for the next week is held, schedule compliance, performance gains or losses, backlog reduction, and so on are reported to the assembled group.

MAINTENANCE NEEDS

Maintenance is primarily concerned with the day-to-day control of work:

- Which new jobs must be acknowledged?
- How will work be requested or identified?
- Once identified, how will work be classified to establish the reaction maintenance must take?
- Is the work an emergency repair, requiring a quick reaction?
- Will work require detailed planning?

Selected jobs become the weekly schedule from which supervisors make daily assignments to crew members.

SUMMARY

An information system will yield performance improvements only if effectively implemented and utilized. Effective maintenance management is linked to quality information provided to well-trained personnel operating within a clearly defined maintenance program. Education is a key factor in successful system implementation. Effective use of information will depend on how well the users understand the way it can help them to perform their jobs. If these points are emphasized in the system implementation, it will be more effective. Work orders provide for requesting, classifying, planning, scheduling, assigning, controlling, and measuring maintenance work. Data provided on work orders feeds the information system where it is converted into useful information. There should be a work order element for each type of job consistent with the need for detail or simplicity.

CHAPTER 13

Essential Information

INFORMATION NEEDS

Different maintenance organization levels will require specific information that will permit them to make informed decisions. Maintenance managers, for example, are interested in summaries and projections, such as

- A monthly absentee summary for the maintenance department,
- Compliance with the vacation policy last month,
- Total maintenance costs year to date, and
- A forecast of major component replacements for 6 months.

Maintenance engineers are interested in technical equipment performance details such as the following examples:

- The life-span of engines in Fleet A,
- The trend of lube failures on conveyor rollers in the past 6 months, and
- The specific repairs made on unit 16 in the past 12 months.

The general supervisor's information focus would include the following types of examples:

- The percentage completion of an overhaul as of the current week,
- Compliance with the preventive maintenance (PM) program last week,
- PM services not completed in area B this week,
- The percentage improvement in worker productivity in the past 6 months,
- The percentage of man-hours able to be used on nonmaintenance work,
- The current backlog of all crafts, and
- A report detailing the cost for each unit in a cost center.

The planner focuses on the preparation of major jobs and obtaining the resources to carry them out. Typical examples include

- The cost of replacing a specific major component;

- The availability of a specific major assembly that is being replaced;
- The man-hours available, by craft, for planned work next week; and
- The delivery status of a major assembly required for an overhaul.

The supervisor is concerned with the current shift, primarily with continuing interest in getting ready for tomorrow. Therefore, his or her information interests span today and tomorrow with examples such as

- The current cost and performance status of a major job,
- A labor use summary showing the use of labor by type of work,
- Mechanics awarded overtime in area B this week, and
- The dramatic rise in the electrician's backlog since last quarter.

Individual craftsmen have information needs, as well, including

- A standard bill of materials for a specific, repetitive job;
- A list of interchangeable bearings; and
- The utilization of specific spare parts.

MAINTENANCE INFORMATION

There are two categories of maintenance information: decision-making and administrative. Decision-making information is used to manage the overall maintenance effort, whereas administrative information is used to manage internal maintenance functions like absenteeism or overtime distribution. Decision-making information includes control of labor, backlog, work status, cost, repair history, and selected performance indices such as worker productivity or schedule compliance (Figure 13-1).

LABOR UTILIZATION

The control of labor is an especially important element of decision-making information. The maintenance work force exists essentially to install materials in the act of making repairs. But the only way that maintenance can control the cost of the work they do is the efficiency with which they install materials. This efficiency, in turn, is brought about by good preventive maintenance, which results in the ability to plan more work and subsequently perform more deliberate, quality work by installing materials efficiently. The result is greater equipment reliability, lower costs, and less downtime. The effective use of labor by category of work is an excellent barometer of the quality of labor control. When adequate man-hours are spent on preventive maintenance, the result is a reduction in labor used on emergency and unscheduled repairs and an increase in labor used on planned and scheduled maintenance. Planned work is better organized and work is performed more productively (Figure 13-2).

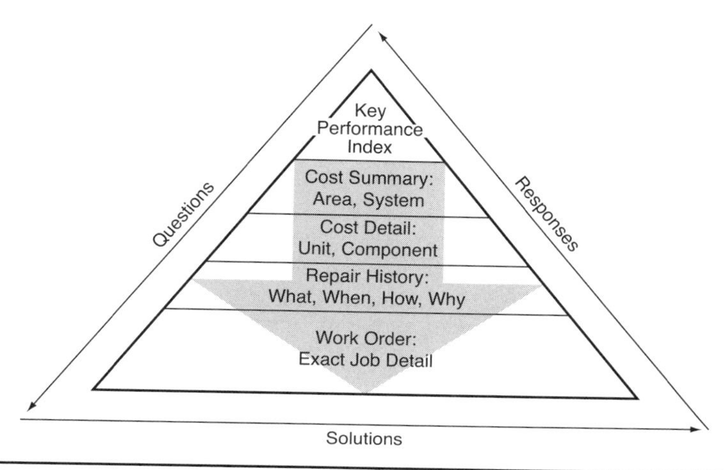

FIGURE 13-1 Information systems are constructed so that a manager wishing to get an explanation of a questionable key performance index can question maintenance. In turn, maintenance can determine the offending cost area, then identify the specific unit of equipment, and pinpoint what was done in repair history. If necessary, exact job detail can be examined. The structure of the system then permits maintenance to respond to questions, develop solutions, and state responses.

MAINTENANCE LABOR UTILIZATION REPORT
WEEK 40 ENDING SEPT. 30

DEPARTMENT 137

CRAFT	PREV MTCE	SCHD MTCE	UNSC RPRS	EMER RPRS	NON-MTCE	TOTL WKFC
MILLWRIGHT	55	276	102	88	67	588
MECHANIC	46	244	96	66	42	494
ELECTRICIAN	23	122	54	32	46	277
WELDER		72	58	21	66	217
INSTRUMENT	18	32	30	12	45	137
PIPEFITTER		48	12	9	12	81
LABORER		121	6	5	41	173
TOTAL MAN-HOURS	142	915	358	233	319	1967
% DISTRIBUTION	7	47	18	12	16	100

FIGURE 13-2 Man-hours reported against categories of work yield a picture of the effectiveness of labor utilization. In this illustration, 7% of labor spent on preventive maintenance has resulted in 47% of labor being used on scheduled maintenance, while holding labor spent on unscheduled and emergency repairs to 18% and 12%, respectively.

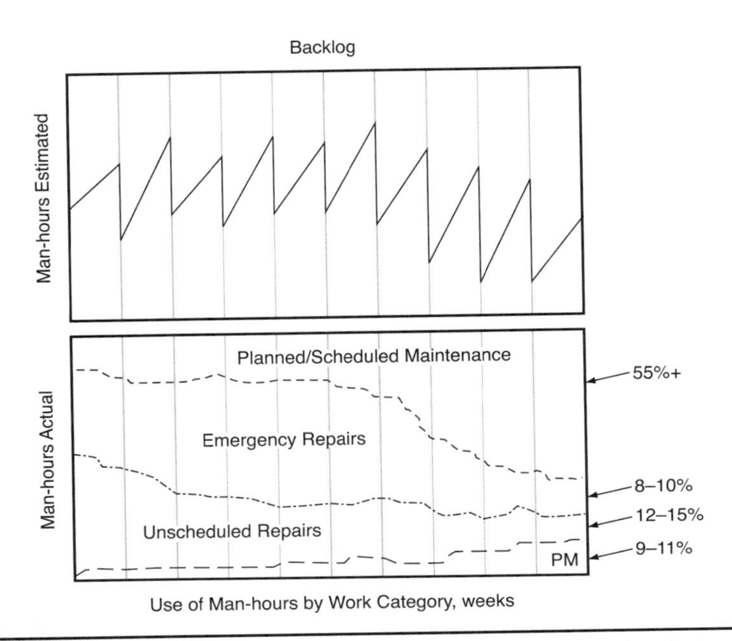

FIGURE 13-3 Target distribution of man-hours utilization by work category. When 55+% man-hours are being used on planned/scheduled maintenance and 9%–11% used on preventive maintenance, most work is being effectively managed. Emergency repairs will never be eliminated, but they can be contained with a strong detection-oriented PM effort. As more man-hours are spent on preventive maintenance, more man-hours will be used on planned/scheduled maintenance, and the backlog will be reduced and controlled.

Labor utilization and the backlog work in concert. As new work orders are opened, the backlog rises by the number of estimated man-hours added. Then as actual man-hours are worked on planned and scheduled maintenance, the backlog is reduced. As more effective preventive maintenance is carried out, more planned work is possible and the backlog is further reduced (Figure 13-3).

IMPROVING PRODUCTIVITY

Productivity measures the quality of labor control. Indices that allow maintenance to observe trends in productivity should be provided. Typically, information on productivity trends might include

- Labor cost to install each dollar of material,
- Man-hours to achieve each utilized operating hour, and
- Man-hours of preventive maintenance for each man-hour of planned work.

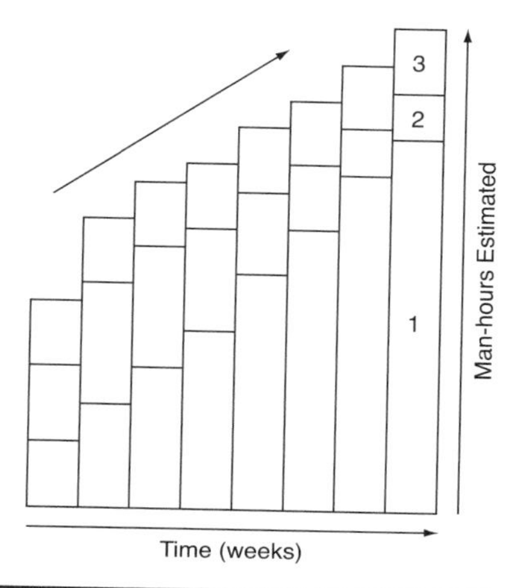

FIGURE 13-4 Backlog data graphed over an 8-week period reveals an overall increase. However, backlog data suggests a need to add craft 1, reduce craft 2, and make no changes to craft 3.

Labor control starts with finding out how many personnel are needed and then ensuring they are used effectively. Effective labor use is the result of good preventive maintenance, effective planning, and quality work control. The ingredient that binds all of these actions together is quality information.

Backlog

The backlog is the number of estimated man-hours by craft required to complete all identified, though incomplete, planned and scheduled work. It is an index of the degree to which maintenance is keeping up with the generation of work as well as a means of determining the proper size and composition of the work force. The backlog includes only the estimated man-hours by craft of all planned and scheduled work. It excludes

- Emergency work because it must be done immediately,
- PM services whose man-hour requirements are the same week after week, and
- Unscheduled repairs too small to require man-hours estimates.

When the backlog is expressed in estimated man-hours by craft, it has specific meaning in helping to adjust the work force size and craft composition. For example, a rising backlog indicates too few personnel, an unchanged backlog indicates there are enough personnel, and a falling backlog indicates a possible excess of personnel for the particular craft. The rising backlog means that more work is being generated than the craft can do. If the rise is long term, more personnel may have to be added. If the rise is shallow and uneven, the use of overtime or the help of a contractor could control it (Figure 13-4).

BACKLOG SUMMARY REPORT
Week: 40 Ending: 30 Sept
MAINTENANCE BACKLOG

ALL PRIORITIES

CRAFT	ST WK	+ WK	- WK	ND WK	ST BL	ND BL	±
MECH	148	13	40	121	148	121	-27
ELEC	96	14	28	82	96	82	-14
WELD	57	10	13	54	57	54	-3
RIGR	128	46	51	123	128	123	-5
FITR	287	49	121	215	287	215	-72
HLPR	147	28	62	113	147	113	-34

FIGURE 13-5 During week 40, the overall backlog dropped by 155 hours. The backlog for mechanics dropped by 27 hours, and the electrical backlog dropped by 14 hours. ST/WK indicates the backlog at the start of the week while +/WK is the number of estimated man-hours added as new work orders are opened. –/Wk is the number of man-hours worked against the backlog adjusted to the current productivity level. ND/WK is the resulting backlog at the end of the week. ST/BL is the total backlog at the start of the week, ND/BL is the backlog at the end of the week, and ± is the net change.

Before work force changes are made as a result of backlog data, using contractors or authorizing overtime should be considered before adding additional personnel. The backlog exists for individual planned work orders. When a work order is opened, the estimated man-hours by craft are added to the backlog. As actual man-hours are worked against the work order, the backlog is reduced. When the work order is closed, any remaining estimated man-hours are canceled. The backlog of work is like a tank of water that fills as the work order is opened, empties partially as the work is done, and empties completely when the work order is closed.

Typically, backlog information is presented in a format that shows changes in the backlog by craft as new work is generated and older work completed (Figure 13-5).

BACKLOG VERSUS PRIORITY

Effective priority-setting is necessary for meaningful backlog data. The priority scheme should meet specific criteria:

- Priorities should be set jointly by maintenance and operations.
- Each rating should have a definite meaning.
- Work of a certain priority should be completed within a specific period.

- The resulting priority numbers should be used to establish the order in which work will be done.
- How labor will be allocated should be specified.

The proper backlog level can be estimated when all work orders move into and out of the backlog within a 16-week period. When a work order has not been completed within 16 weeks,

- The work may have been done already, piecemeal;
- The work was given too high a priority; or
- The equipment may no longer exist.

Placing a job in the backlog does not mean that further deterioration of the equipment ceases. Rather, as it continues to operate, it deteriorates further. Often, operations personnel, concerned that it may fail, ask for help in making repairs that are not initiated on the same work order. Thus, the original work order, lingering in the backlog, does not reflect the true condition of the equipment. Backlog information helps to determine whether maintenance is keeping up with the generation of new work, and it provides a means of adjusting the size and composition of the work force as the work load changes.

BACKLOG MATHEMATICS

Three relationships exist between the estimated and actual man-hours as the work order is completed:

- Actual man-hours are more than estimated man-hours.
- Estimated man-hours are more than actual man-hours.
- Actual man-hours are equal to estimated man-hours.

The backlog computation rules on closing the work order are

- Actual man-hours are more than estimated, and the backlog = 0.
- Estimated man-hours are more than actual hours, and the remaining estimated hours are canceled.
- There is no negative backlog.

WORK ORDER STATUS REPORT

The work order status report allows the planner to observe individual planned/scheduled maintenance jobs from inception to completion and, upon completion, to determine the cost and performance of the completed job (Figure 13-6).

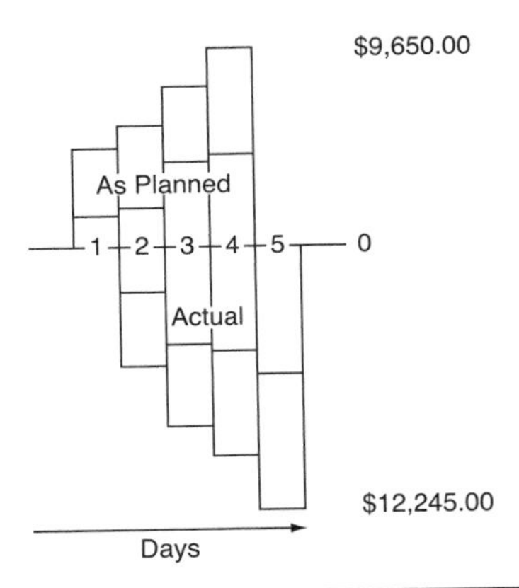

FIGURE 13-6 This work order example started one day later than planned, and its actual cost exceeded the estimate. Data showing the actual status of the individual job would be shown in the work order status report.

MAINTENANCE WORK ORDER STATUS REPORT

MWO#	Sec	Unit	Comp	Ct	Unit Name	Comp Name
426820	024	CM01	100	S	Continuous Miner	Cutting

Description of Work: RPL CUTTER HEAD MOTOR +

St	Opn	Stt	Cls	Pr	Wa	FC	RH	SN/In	SN/Out
S	0610	0612		90					

Work Order Notes

See Material List

	EST	ACT
TR1	12	
TR2		
TR3		
LABOR	300	
MATERIAL	25600	
TOTAL	25900	

Completed:

Requested:
B. Easton
04/03/09

Approved:
J. Southworth
04/16/09

FIGURE 13-7 The work order status report shows the detail at the time the job is opened, and a link is made with the accounting system via the MWO number

As the planner opens the computer screen to reveal a specific work order status report, she makes a link between the maintenance work order (MWO) number, the job, the equipment on which the work will be carried out, job details such as estimated man-hours, and the accounting system (Figure 13-7).

As planning continues, the planner uses the purchase order to obtain direct charge materials and the stock issue card to obtain stock parts for the

MAINTENANCE WORK ORDER STATUS REPORT

MWO#	Sec	Unit	Comp	Ct	Unit Name	Comp Name
426820	024	CM01	100	S	Continuous Miner	Cutting

Description of Work: RPL CUTTER HEAD MOTOR +

St	Opn	Stt	Cls	Pr	Wa	FC	RH	SN/In	SN/Out
S	0610	0612		90	No			128875	122889

Work Order Notes

See Material List

	EST	ACT
TR1	12	8
TR2		
TR3		
LABOR	300	200
MATERIAL	25600	21000
TOTAL	25900	21200

Requested:	Approved:	Completed:
B. Easton	J. Southworth	
04/03/09	04/16/09	

FIGURE 13-8 A work order status in progress shows accumulated material and labor costs as well as actual man-hours. These are compared with the planner's estimates.

job. By placing the work order number on each of these documents, the cost of the materials is debited against the job. When the availability of materials is confirmed, the planner places the MWO on the schedule, obtains approval of the schedule at the weekly scheduling meeting, and sends the approved schedule to the supervisors. Each MWO can be seen on the computer by any user. Work has now begun and the status of the MWO can be seen in the next stage, as shown in Figure 13-8.

As the work is scheduled and performed, each craftsman places the work order number on his or her time card opposite the hours worked to accumulate actual man-hours. In turn, the information system summarizes these costs and related data and debits them against the job. As the job is completed, the supervisor adds details such as

- Whether the job satisfies a warranty (WA),
- The failure code (FC),
- If the job is to be placed in repair history (RH),
- The serial number of the component installed (SN/In), and
- The serial number of the component removed (SN/Out).

When the supervisor signals the planner that the work is completed, the work order may be closed. Then cost and performance details are summarized. See Figure 13-9.

MAINTENANCE WORK ORDER STATUS REPORT

MWO#	Sec	Unit	Comp	Ct	Unit Name	Comp Name
426820	024	CM01	100	S	Continuous Miner	Cutting

Description of Work: RPL CUTTER HEAD MOTOR +

St	Opn	Stt	Cls	Pr	Wa	FC	RH	SN/In	SN/Out
S	0610	0612		90	No			128875	122889

Work Order Notes

See Material List

	EST	ACT
TR1	12	
TR2		
TR3		
LABOR	300	300
MATERIAL	25600	25720
TOTAL	25900	26020

Requested:	Approved:	Completed:
B. Easton	J. Southworth	T. Washburn
04/03/09	04/16/09	04/18/09

FIGURE 13-9 This work order status report shows that the work order has been closed. The cost and performance details are summarized and compared with the planner's original estimates.

Because the work order (MWO) has isolated the cost and performance of a major planned and scheduled job, closing performs two functions:

1. The costs of labor and material are closed against the unit of equipment.
2. The job may be recorded in repair history to establish a pattern of repairs to help confirm problems and determine corrective actions.

The planner can administer numerous work orders in various stages. Some are ready to schedule, while others await materials. The various statuses of pending work orders might include

- Waiting for materials
- Waiting for labor
- Waiting for shutdown
- Waiting for completion of shop work
- Ready
- Scheduled or deferred
- Completed

Generally, it is not advisable to spend too much time comparing the estimated and actual costs, man-hours, or timing on single completed work orders.

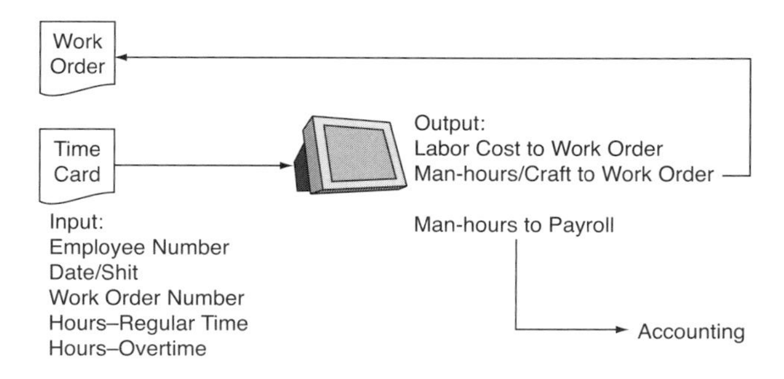

FIGURE 13-10 As labor is reported against the work order, its detail provides both labor cost and man-hours used by craft against the work order. In addition, the actual man-hours are removed from the backlog and also sent to accounting to facilitate payroll computation.

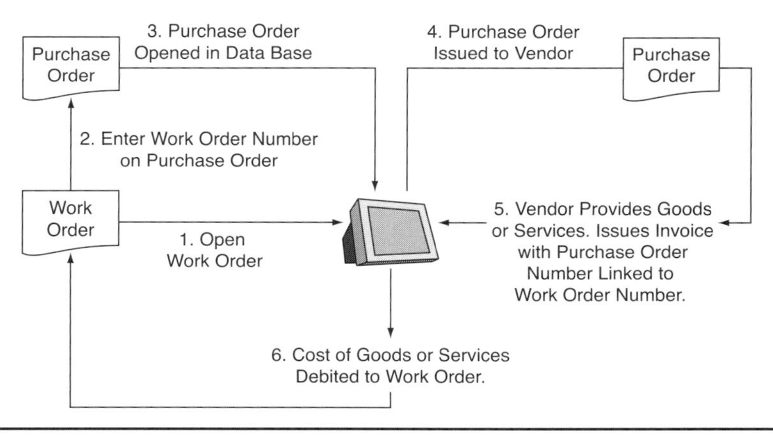

FIGURE 13-11 Purchased materials are debited to the work order via the purchase order

Each weekly schedule should be examined so that goals can be set for better overall performance. For example,

- Set a goal of estimated versus actual hours or cost for all work orders on the schedule of +10%.
- Establish a target schedule compliance of 85%.

Throughout the use of the work order status report, actions are in process between the work order system and the accounting system (Figure 13-10). In addition, material needs are being met concurrently (Figure 13-11)

COST INFORMATION

Cost information exists at two levels. At the first level, a cost summary provides the comparison of actual versus budgeted cost by cost center for all maintenance activities (Figure 13-12). The second level of cost information is the cost detail showing the cost by unit and component permitting a detailed analysis of expenditure (Figure 13-13).

Cost detail summarizes the cost of units and their components on a month- and year-to-date basis. When the cost information is assessed and the offending equipment and components identified, comparisons are made with similar equipment to help determine how poorly the equipment is performing. Then, as necessary, specific corrective actions can be determined (Figure 13-14).

Verification of improved equipment performance can be confirmed with detailed cost reports that examine individual unit performance at the unit and component level. Additional reports contrast maintenance and operations production statistics (Figure 13-15).

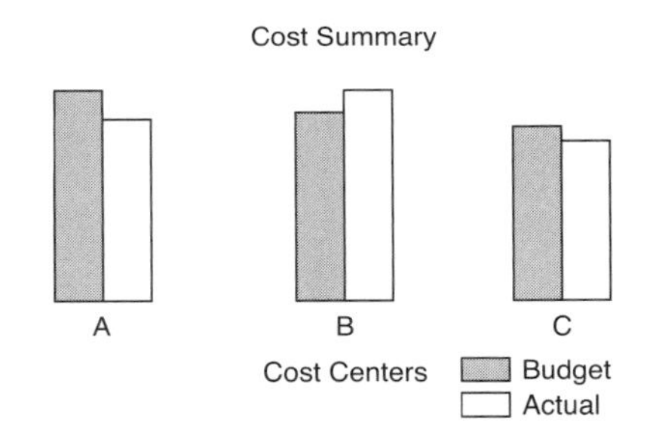

FIGURE 13-12 The actual cost in cost center B exceeds its budget

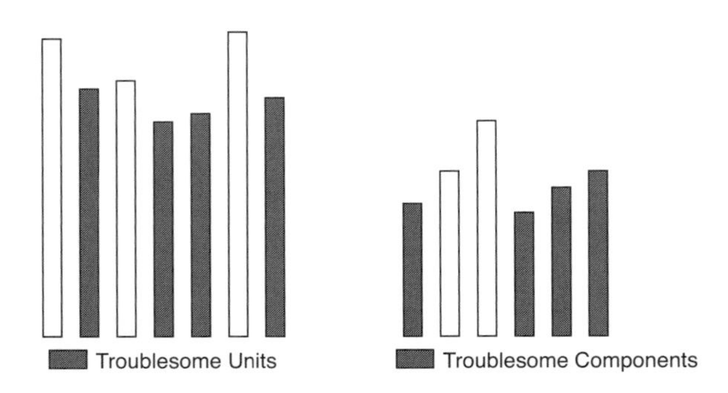

FIGURE 13-13 The cost detail report identifies both troublesome equipment and components experiencing higher than normal maintenance costs

EQUIPMENT COST BY TYPE/COMPONENT
FEBRUARY

TYPE CP	CURRENT MONTH MANHOURS RT	OT	LBR CST	MTL CST	TTL CST	YEAR TO DATE MANHOURS RT	OT	LBR CST	MTL CST	TTL CST	AVG CST MONTH
51 LHDS											
00 GENERAL	1021.5	29.0	10666	10895	21561	2595.0	150.5	28933	25756	54689	27344
01 ENGINE	490.5	39.0	5930	30993	36923	1119.0	84.5	13546	52324	65870	32935
02 CONVERTER	13.0	1.0	217	24	241	66.0	3.0	755	295	1050	525
03 TRANS	358.5	24.5	4198	12732	16930	797.5	55.0	9312	12892	22134	11102
04 DRIVE-TRAIN	592.0	28.5	6668	17294	23962	1291.5	87.0	15131	29702	44833	22416
05 TIRES	132.5	11.5	1543	10634	12177	289.5	26.5	3432	58180	61612	30806
06 BUCKET	479.5	23.0	5504	2800	8304	770.5	30.5	8715	8389	17104	8552
07 HYDRAULICS	625.5	44.5	7331	7166	14497	1664.0	102.0	19438	11477	30915	15457
08 BOOM	43.0	0.0	485	1458	1943	130.0	3.5	2306	4002	6308	3154
09 BRAKES	495.0	39.5	5939	4460	10399	913.5	71.5	11014	7330	18344	9172
10 FRAME	902.0	60.0	10448	10292	13740	1814.5	105.5	21091	15099	36190	18096
12 ELECTRICAL	249.5	15.0	2945	3172	6117	539.0	31.5	6384	8555	14939	7469
13 SERVICE	10.5	0.0	112	0	112	38.5	0.0	405	0	405	132
TOTAL	6411.0	315.5	61996	111913	173916	11396.5	751.0	140462	234001	37446	187231

FIGURE 13-14 This cost report signaled a need to investigate possible fleet-level problems with engines, drive trains, tires, and frames. It also established a need to clarify exactly what would be reported under "general"—one of the largest cost elements.

FLEET COST/PERFORMANCE REPORT
APRIL

PRODUCTION 8 CU.YD. LHD

TYPE/ UNIT	CLNUP HRS	MTC HRS	SVC HRS	IDL HRS	DLAY HRS	SETUP HRS	WORK AVL	% UTL	% W-HR	BKTS HRS	BKTS/ T-HR	TTL	BKTS/
510516	3.3	271.5	7.0	15.5	13.0	0.5	231.8	50.6	87.2	9302	40.1	549.8	16.9

COST/BUCKET OPNS: 0.27 F&L: 0.01 TIRES: 0.56 MAINT: 0.58 TOTAL: 1.42 YTD: 1.42
F & O/PER BUCKET DIESEL: 21 GAL/0.03 10-WT: 0 GAL/0.0 30-WT: 0 QTS/0.0 ATF: 8 GAL/0.0
TRAM INFORMATION MILES: 326.74 FT/BUCKET: 185.55

TYPE/ UNIT	CLNUP HRS	MTC HRS	SVC HRS	IDL HRS	DLAY HRS	SETUP HRS	WORK AVL	% UTL	% W-HR	BKTS HRS	BKTS/ T-HR	TTL	BKTS/
510571	8.0	87.6	15.8	44.8	28.0	0.0	368.5	84.2	84.3	15596	42.3	552.7	28.2

COST/BUCKET OPNS: 0.26 F&L: 0.02 TIRES: 0.10 MAINT: 0.24 TOTAL: 0.61 YTD: 0.761
F & O/PER BUCKET DIESEL: 445 GAL/0.03 10-WT: 41 GAL/0.0 30-WT: 13 QTS/0.0 ATF: 2 GAL/0.0
TRAM INFORMATION MILES: 680.68 FT/BUCKET: 230.4

FIGURE 13-15 Note the cost-related information, including cost per bucket (maintenance), and fuel and oil per bucket (diesel fuel). Other performance elements include maintenance hours, availability, and utilization.

REPAIR HISTORY

Repair history allows the observation of chronic and repetitive problems, confirmation of failure modes, observation of failure trends, determination of failure frequencies, observation of corrective actions taken, determination of whether the problem was corrected, measurement of component life-spans, and the capability to develop a solution and observe testing of that solution (Figure 13-16). A typical repair history report is shown at Figure 13-17.

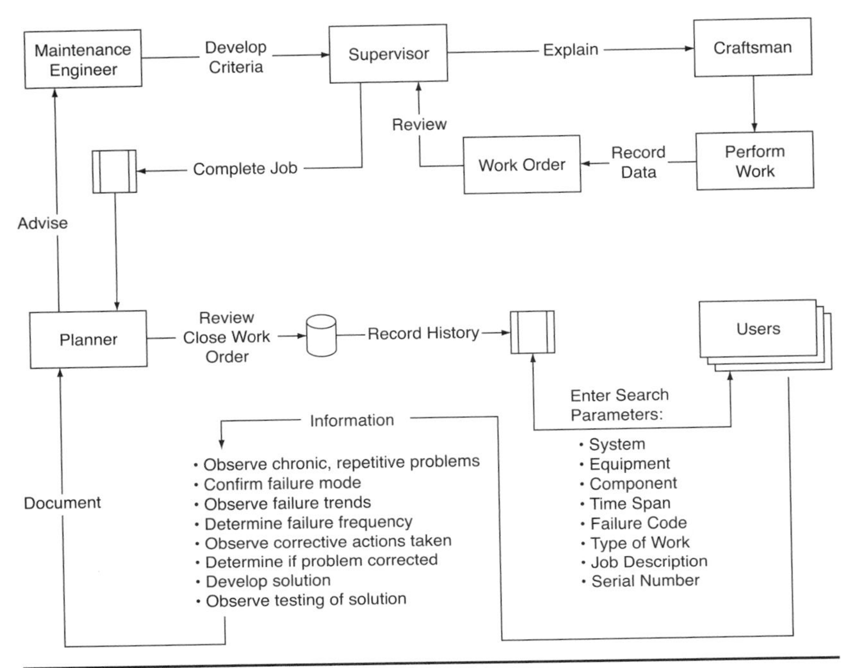

FIGURE 13-16 Flow of repair history

REPAIR HISTORY

FC	MWO#	Sec	Unit	Comp	Tr	MH	YrWk	Description of Work	SN (In)	SN (Out)	OpHrs
1	64236	024	CM02	400	EL	8	0643	RPL TRAM MTR OFF SIDE (JOY)	112123	312413	1102
2	65135	024	CM02	400	EL	6	0652	RPL TRAM MTR OFF SIDE (JOY)	223335	112123	2781
2	65223	024	CM02	400	EL	8	0705	RPL TRAM MTR OFF SIDE (JOY)	124498	223335	1286
2	65465	024	CM02	400	EL	5	0729	RPL TRAM MTR OFF SIDE (TRAM)	344456	124456	1589
2	65553	024	CM02	400	EL	5	0738	RPL TRAM MTR OFF SIDE (TRAM)	129032	344456	1703
2	65664	024	CM02	400	EL	6	0743	RPL TRAM MTR OPP SIDE (JOY)	590234	129032	3307
1	65701	024	CM02	400	EL	8	0748	RPL TRAM MTR OPP SIDE (NAT)	452903	590234	3490

FC = failure code; Tr = trade; YrWk = year/week; OpHrs = operating hours.

FIGURE 13-17 This repair history isolated the actions taken on CM02 (continuous miner 02), component 400 (off side tram motor) and determined that the most frequent cause of failure was improper lubrication (2). It also identified poorly performing tram motors. Serial number 112123, for example, lasted only 10 weeks (0643 to 0652) and only l,679 operating hours (2,781 – 1,102).

ADMINISTRATIVE INFORMATION

Among the types of administrative information used internally by maintenance is the management of overtime. Overtime use should be classified by reason for use to determine how effectively it is being used and controlled. Concurrently, information on overtime confirms whether established overtime approval policies are complied with and overtime is used effectively. See Figure 13-18.

Absenteeism that is poorly controlled can be a major detriment to the effective control of labor. Therefore, the same level of control applied to the use of labor must be applied to absenteeism. Typically, an effective absentee report shows the cause of absenteeism. See Figure 13-19.

EQUIPMENT SPECIFICATION PROGRAM

An equipment specification program allows a craftsman (or other user) to specify equipment type and component, and obtain a listing of corresponding stocked parts or component details (Figure 13-20). The computer responds with the details found in Figure 13-21.

Equipment specification programs permit a visual presentation of components as well (Figure 13-22). The craftsman views an "exploded view" of a component assembly, allowing him or her to select the correct part, review assembly details, and even print a copy of the diagram and take it to the work site if needed. This type of graphic presentation has become very popular and useful.

MAINTENANCE OVERTIME SUMMARY
WEEK: 40 ENDING: 30 SEPT

OCC	1	2	3	4	5		
CODE	DESCRIPTION	CALL-IN	EMERGENCY	SCHEDULED	UNAUTHORIZED	CONTRACT	TOTAL
01	RIGGER					5.0	5.0
03	DIESEL RPRMAN	10.0					10.0
04	WELDER/ MECHANIC			13.0			13.0
05	MECHANIC			35.5			35.5
10	FITTER						3.0
17	MACHINIST		6.0				6.0
13	ELECTRICIAN		2.0				2.0
25	WELDER				12.0		12.0
37	BOILER MAKER		1.0				1.0
TOTAL OVERTIME MH		98.0	46.0	80.0	14.5	18.0	265.5
% DISTRIBUTION		38.2%	17.9%	31.1%	5.7%	7.1%	100.0

FIGURE 13-18 A weekly report of overtime use by craft and reason helps to determine how effectively overtime is being used and controlled

MAINTENANCE ABSENTEE SUMMARY WEEK: 40 ENDING: SEPT 30												
	PER	UNN	VAC	JRY	LVV	INJ	NAT	BRV	LTE	DSP	OTH	TOT
DESCRIPTION	BUS	BUS	ION	DTY	ABS	URY	GRD	MNT	RPT	ACT	ER	AL
MILLWRIGHT	12	6	80		8	2						108
MECHANIC	4	2	40	2	8			8				64
WELDER	1	3	40			8			2		1	55
MACHINIST	4								4			8
ELECTRICIAN	5	8	80		4	1			2	2	1	103
INSTRUMENT		1					40		1	1	1	44
PIPEFITTER	1	4	40		3	4			2		1	55
TOTAL MH	27	24	280	2	23	15	40	8	11	3	4	437
% DISTRIBUTION	6	5	64	1	5	3	9	2	3	1	1	100

FIGURE 13-19 The absentee report allows the reader to assess the effectiveness with which absenteeism is being controlled

```
(INQUIRY)          Equipment ID - CV010-130              09/08
                   List of Components
-----------------------------------------------------------------
011                Electric motors                019    Idler shafts
015                Reduction units                013    Drive belts
018                Sprockets (roller chains)      021    Transfer belts
```

FIGURE 13-20 The user inquires by equipment identification, CV010-130 (coal conveyor #10, component 130) requesting data on the parts that make up component 130; in this instance, 011 (electric motors).

```
(RESPONSE) CV010-130          #10 Coal Conveyor              09/08
                         011 Electric motors
-----------------------------------------------------------------
MOTOR ITEM # ( )

MAKE: CBE            MODEL: 9F3266N          HORSEPOWER 50

RPM: 1760            FRAME: 254U             VOLTAGE: 440    AMPS: 13.5

OTPT SHAFT DIA: 1.375"        DE BEARING #: 40BC0J

ODE BEARING #: 358C0J
-----------------------------------------------------------------
```

FIGURE 13-21 The equipment specification program responds, for the component selected, with details of the motor: horsepower, frame, revolutions per minute, etc.

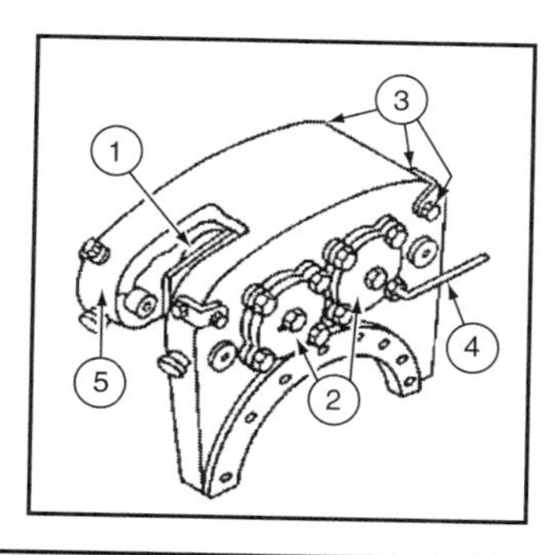

FIGURE 13-22 Visual presentation of components

ASSESSING MAINTENANCE PERFORMANCE

The ultimate objective of good maintenance performance is to ensure plant profitability. In turn, achievement of this objective depends on quality information shared between management, operations, and maintenance. Generally,

- Management needs trends and summaries to assess the overall effectiveness of maintenance;
- Operations needs assurance that their operating funds are well-spent and quality work is being done; and
- Maintenance needs to identify work, then control the work and measure how well they have accomplished it.

Performance indices and trends are the principal means by which management can assess performance. Among them,

- The unit of product per maintenance dollar reveals the overall contribution of maintenance in the plant productive effort.
- The maintenance cost per operating hour assesses how well maintenance contributes to the reduction of downtime.
- The labor cost to install each dollar of material reveals how effectively maintenance uses its labor to perform work.

An unsatisfactory performance revealed by a declining trend must be followed by a detailed investigation to identify the problems and make corrections. Management uses indices to require explanation by maintenance and followup corrective actions. As corrective actions are taken, management continues to observe the indices and note the impact of the corrective actions on future trends. With these types of indices, managers can observe the trends that indices point out and, with confidence, ask maintenance to identify and deliver corrective actions when the indices reveal unfavorable trends.

SPECIFIC MAINTENANCE PERFORMANCE

In addition to trends, other management indices should identify specific maintenance performance, such as

- Compliance with the PM schedule, and
- Compliance with the weekly schedule of major planned and scheduled work.

These indices reveal whether the maintenance program is being carried out effectively and provides management focus for directing corrective actions. For example,

> A plant manager was concerned that compliance with the PM schedule was declining. On attending the weekly scheduling meeting, he learned that poor compliance was attributable to operations not making equipment available on time. Thus, his corrective actions would be based on more careful delineation of operations responsibilities than criticism of maintenance performance.

MAINTENANCE COSTS

Management is also interested in actual maintenance costs versus budgeted costs on a current month and year-to-date basis. This assessment reveals the overall quality of maintenance management. Although trends will reveal a stable, improving, or declining performance, they do not pinpoint problems.

PERFORMANCE INFORMATION FOR OPERATIONS

A primary interest of operations will be the cost of the maintenance service it receives. However, operations personnel will not want to be inundated with details. A cost summary showing actual cost versus budgeted costs for each cost center is an excellent starting point. This information will alert operations to troublesome areas and open the way for suggesting solutions. After the troublesome areas are identified and discussion has begun, confirmation of problems is in order. In addition, a weekly listing of all repairs detailed by department, unit, and component made available on line is not only greatly appreciated and widely used by operations personnel, but it assures them that they are being well-served.

SHARING INFORMATION

Maintenance can confirm which units and components are at fault for their high costs. In addition, the nature of what happened can be revealed by repair history. However, little will happen unless joint action results from use of the information. For instance,

- If operator error results in excessive damage to equipment, high maintenance costs will not be reduced until training produces more competent operators. Thus, operations should scan the failure summary contained in the repair history.
- Similarly, if operations fail to make equipment available for scheduled servicing or repairs, they must be held accountable for excessive maintenance costs. Therefore, it is in the interest of operations personnel to note PM compliance and, if poor compliance is their fault, act to improve it.

DOWNTIME

Downtime is nonproductive time for equipment, which is expensive and unrecoverable. Its causes must be identified and reduced. Although inadequate maintenance is a contributing factor, operations contribute as well. Too often, causes of downtime are incorrectly attributed to maintenance because operations is usually the "scorekeeper" (it prepares the downtime report to explain why production or quality targets are not met). Therefore, downtime performance data should be identified and implemented jointly by maintenance and operations. Maintenance should be allowed to countersign the downtime data submitted. It is their scorecard as well. Generally, downtime is attributable to maintenance when

- Maintenance has been notified and is still responding. This measures the effectiveness of maintenance communications.
- Maintenance is performing the work to correct the problem. This measures the efficiency with which maintenance gets its work done.
- Equipment is down while materials are being procured. This measures the quality of material control.

Downtime is attributable to operations when

- Equipment operators are absent or late.
- There are production delays.
- Equipment is not put back into service promptly on completion of repairs.

Questionable areas must be resolved face to face, such as, for example, when operations ignores a PM service and the equipment fails.

REFINEMENTS

After the maintenance program is in place, refinements relating to the way in which the maintenance services are carried out will be important. For example,

- Compliance with the PM schedule indicates concern in ensuring reduction of emergency repairs and downtime. Management, operations, and maintenance should observe this trend.
- Compliance with the weekly schedule indicates interest in the effectiveness with which maintenance has accomplished major jobs. Again, all three parties should observe it.

INTEREST IN QUALITY

A sure sign of operations' interest in the maintenance program is its demand for quality work. There are a number of ways it can be achieved. Consider the following scenario:

> At a remote timber operation, the "bush" operation did minimal equipment servicing and were dependent on a distant garage for all major servicing and repairs. Because equipment operators were on an incentive program, they became increasingly critical (often vocally) of the quality of work being performed by the garage. Supervisors worked long and hard to solve the problem with little progress. Finally, operators volunteered to travel to the garage and road-test each unit before it was transported back to the bush. The first few tests were confrontational. But then the operators and the mechanics started to communicate, and the complaints soon disappeared. Management marveled at the result of a little bit of direct quality control. The customer and the service provider better appreciated each other's problems and needs.

Typically, a face-to-face quality effort brings good results:

- In a mineral processing plant, operations recommended that maintenance craftsmen have operators inspect completed repairs. The results, measured in quality work, were dramatic. For the first time, craftsmen got the satisfaction of being told directly that they had done a good job.
- In a foundry, maintenance supervisors asked production supervisors to test-run equipment before the maintenance crew left the work site. Maintenance crews took this opportunity to point out that equipment cleanup before work began could help them. The result was better overall cooperation and better cleanup, illustrating that quality control can work both ways.

PERFORMANCE INFORMATION HELPFUL TO MAINTENANCE

Maintenance is ultimately judged on its ability to minimize downtime, reduce cost, and perform quality work. To do this, they must identify the work, control the work, and measure how well they accomplished it. These needs span both the work order and information systems. Thus, to identify work quickly and accurately, maintenance must

- Have an easy-to-access work request system,
- Be able to respond to verbal requests,
- Have a strong detection-oriented PM program,
- Convince operations to report problems promptly,
- Have continuous information on the status of jobs,
- Obtain timely cost information, and
- Get up-to-date information on the use of labor.

To measure the effectiveness of work accomplishment, maintenance must have

- Knowledge of job cost and performance,
- Information on worker productivity,
- The ability to compare actual and estimated job costs, and
- Access to actual versus budgeted maintenance costs.

Thus, the success of any maintenance information system is the result of effective conceptualization, development, implementation, and information utilization by those who manage the plant, those who receive maintenance services (operations), and by maintenance as they perform the service. The successful improvement of maintenance cost and performance relates directly to the availability of information and its effective use by management, operations, and maintenance. Therefore, the design of the maintenance information system must acknowledge these information needs, and system implementation must deliver the needed information. It follows that education on information use must be built into the implementation and utilization phases.

THE RIGHT PERFORMANCE INDICES

There are many indices of maintenance performance, including labor use, downtime, cost, backlog reduction, and so forth. Each measures a specific element. However, no single index of performance produces a total picture. Rather, a combination of indices is required to yield an overview of maintenance performance. Generally, performance indices are aimed at providing decision makers with a comparison of current performance against trends or desired levels of performance. Thus, the indices must trace performance in a historical sense. In addition, the indices must be derived from readily obtainable information or data. This permits maintenance to seek answers to an improvement need suggested by a poor current index in its own information system. For example, if the cost-per-ton index reveals an unsatisfactory ratio, maintenance should be able to find the offending equipment in its cost reports and then isolate the problems in existing repair history. Thus, they can respond quickly with corrective actions or solutions.

FAMILY OF INDICES

Each plant must develop a family of indices that best portrays their overall performance picture. The plant must consider several factors:

- Different management levels require different information,
- Which indices best describe actual maintenance performance, and
- How easily supporting data can be obtained.

USE OF INDICES

The information needed by the plant manager versus the maintenance superintendent or the maintenance supervisor differs in both its nature and intent. The plant manager, for example, needs summaries and trend information that tell him the plant is headed in a safe direction. Yet, his time must be divided among a multitude of interests and concerns. Thus, he has little time to study reports. But he must be well-informed. Consequently, he has the greatest need for indices that give him a comprehensive performance picture quickly. By contrast, the supervisor wants immediate information. She must know what happened on the previous shift and what new problems were created. She has less interest in trends and summaries (Figure 13-23).

COST VERSUS PERFORMANCE

The cost of plant operation and the contribution of maintenance to that cost are of concern to a plant manager. Managers are intent to push costs down to levels that will ensure plant profitability. The most useful indices should appreciate the forces that drive maintenance costs. Consider the following factors:

- Total maintenance costs are controlled by the cost of material used rather than by the cost of labor. Most maintenance departments have a stable work force; thus, total labor cost variations are largely due to overtime.
- Maintenance costs are determined by the volume of equipment that must be maintained and the speed at which equipment repair consumes spare parts and materials. Thus, equipment repairs characterized by repeated emergencies consume materials faster than equipment subjected to a deliberate, well-ordered maintenance program. Therefore, a key maintenance cost-reduction objective should be to reduce emergency repairs to slow the rate at which materials are consumed.
- Labor costs are incurred by the need to install materials. But when emergency repairs are done less productively, more labor is consumed

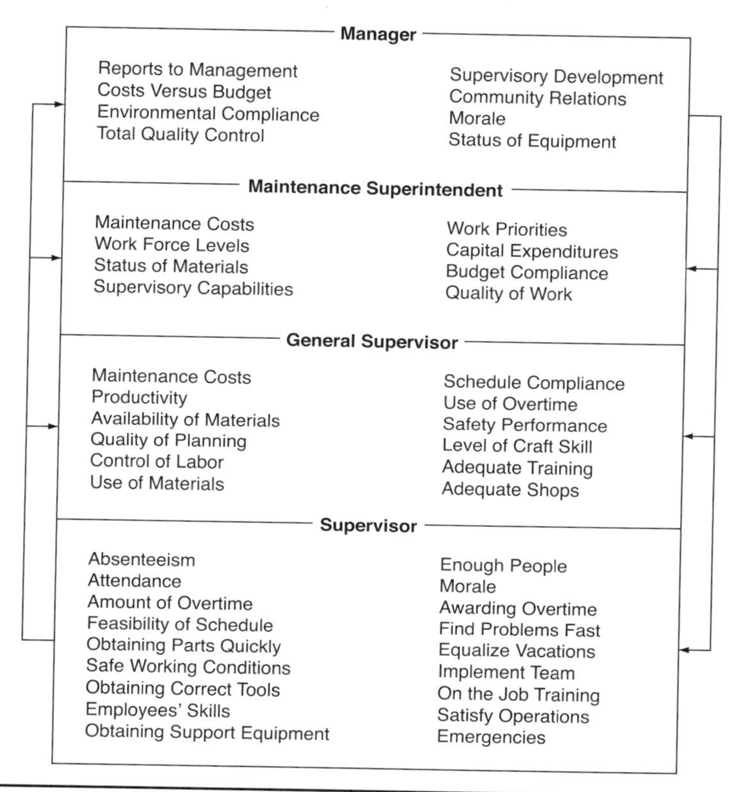

FIGURE 13-23 Information versus management levels. The plant manager has the greatest need for meaningful performance indices which give a total performance picture in the least amount of time. First-line supervisors, on the other hand, require indices that inform them of the accomplishment of day-to-day work.

to install the same materials. The work is often rushed, resulting in lower work quality. Often, emergency repairs are done simply to get the equipment running again. Repairs must be repeated when equipment is less urgently needed. Thus, the expenditure of labor on emergencies is not only inefficient but often it may be wasteful because the work must be redone.

- The only significant control that maintenance has over the cost of the work they perform is the efficiency with which they install materials. Thus, better preventive maintenance, leading to fewer emergencies and more planning, results in better labor utilization and less labor cost to install the materials.

- As work is carried out, maintenance exercises better control over individual jobs that are planned and scheduled. These jobs, because they are better organized, are often completed in less elapsed downtime, reducing the overall cost because less labor is used. Generally, material costs are the same whether the job is planned or not. More planned work also reduces downtime.
- Maintenance supervisors make the greatest contribution to cost control through the productivity increases they bring about with quality supervision.

Based on these factors, the control of labor then becomes a vital maintenance function, and indices that reflect how efficiently it is used should be included in the overall performance picture.

OVERALL INDEX OF MAINTENANCE PERFORMANCE

Perhaps one of the most positive overall indices of maintenance performance is the cost of labor to install each dollar of material. If applied at the department level, it provides a macro picture. But it can be applied at the crew level as well, revealing the relative productivity of the crew. Based on the average 50% split of maintenance labor and material costs, an efficient maintenance department should be able to install each dollar of material with between $0.87 and $1.02 of labor. Labor costs in excess of this range signal poor preventive maintenance, little planning, ineffective work control, and unsatisfactory control of labor, as shown in Figure 13-24.

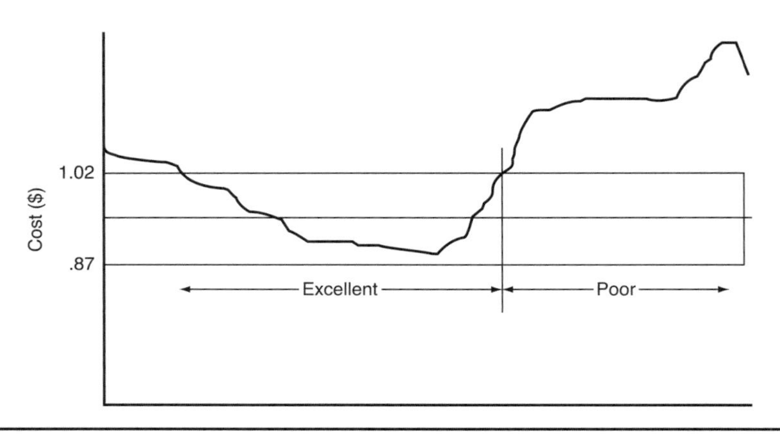

FIGURE 13-24 Cost of labor to install each dollar of material

RELATIONSHIP BETWEEN INDICES

A close relationship exists between operating costs, major repairs, and total maintenance costs, as is apparent in the following example:

> One plant focused on the management of its labor. Their findings revealed that labor cost, as a percentage of total cost, was a good long-term measure of labor efficiency. By upgrading preventive maintenance and improving planning, the plant was able to reduce the ratio of labor use from 48% to 41% of total maintenance costs. They realized a cost savings of more than $95,000 per month.

Thus, there is little question that labor data must be included in performance indices. However, a broader spectrum of indices is necessary to gain the overall picture of maintenance performance. In developing these indices, it is better to focus on the basics that ensure better performance than to identify a poor performance but not know how to improve it.

USEFUL INDICES

It is better to check the compliance with the PM program than to measure productivity. If PM services are done on time, it will permit more planned work and ensure better productivity, thus eliminating the need to check short-time productivity. Measuring performance on single major jobs is not particularly useful if compliance with the total schedule is greater than 85% and all jobs were completed within ±10% estimated versus actual cost. Each index should be meaningful.

Likewise, total maintenance downtime is rarely useful except to use maintenance as a scapegoat. But if downtime is divided into meaningful segments, it can bring about real improvements. For example, the downtime awaiting maintenance measures the effectiveness of internal maintenance communications. Downtime spent performing work reveals the efficiency of planning and control, whereas the downtime awaiting materials states the quality of material control support. Mechanical downtime versus electrical downtime is void of any useful information.

LINKAGES OF INDICES

Many indices are interrelated. If, for instance, the maintenance supervisor spends little time on effective supervision, productivity will be poor. Generally, at least 60% of the supervisor's time must be spent on supervision to yield a minimum of 40% productivity, which for an 8-hour shift is only 3.2 hours of productive work. The plant manager and the maintenance superintendent will utilize different indices with different objectives. The plant manager is interested in profitability and meeting production targets:

- Total maintenance cost
- Budget compliance
- Capital expenditures on new equipment
- Percent of total operating cost
- Cost per ton of product by plant
- Cost per operating hour per fleet

The maintenance superintendent, on the other hand, is interested in the actual performance of her department. Thus, her need for performance indices is more extensive and it covers areas that she must manage personally. Moreover, her indices must indicate exact problems, enabling her to identify the supervisor responsible for improving it. In turn, that supervisor must be able to pinpoint the problem using available information. The following lists portray the broad but direct areas that the maintenance superintendent must be concerned with regarding labor control:

- Productivity
- Percent of labor used on emergency repairs
- Percent of labor used on planned maintenance
- Backlog level
- Percentage of overtime
- Compliance with vacation schedule
- Compliance with absentee policies

Control measures:

- PM compliance
- Compliance with weekly schedule
- Percent supervision time

Costs:

- Labor cost to install each dollar of material
- Cost per productive hour
- Percentage of maintenance to total operating cost
- Cost per ton in each plant
- Cost per operating hour by fleet
- Total maintenance cost
- Cost of overtime

Downtime:

- To respond to operations requests
- To perform work
- Awaiting materials

SUMMARY

Quality information impacts performance through better control and effective decision making. Although information systems have made strides toward better data handling, maintenance personnel must take greater advantage of the better information capabilities available. In turn, performance indices are tied to information specifics so that corrective measures can be quickly identified and applied.

CHAPTER 14

Nonmaintenance Project Work

Nonmaintenance projects include construction, equipment modification, new equipment installation, and the relocation of equipment to a new plant location to perform the same function. Most nonmaintenance work creates a new entity, as with newly installed equipment or a recently constructed building. Capital funds must be approved to finance the work. In process plants, equipment modification requires an engineering review of the modification impact. Any new major project should have initial management sanction, and policies should ensure the necessity, feasibility, and funding source. Often, production managers may not distinguish nonmaintenance from maintenance. They could innocently levy project requirements on maintenance for equipment modifications just as if they were regular maintenance. Therefore, screening by management and engineering is necessary before the work is assigned to maintenance. See Figure 14-1.

Maintenance should avoid getting overly involved in project work at the expense of not performing basic maintenance.

> In one instance, a plant found to be doing excessive project work and too little maintenance was discovered to have a work force made up more of "frustrated contractor type" craftsmen than "maintenance" craftsmen. A review of the work completed revealed that more than 30% was solicited by them and their "contractor-oriented" supervisors. On making this discovery, the plant manager learned why maintenance was so poor but was disappointed to find that projects were being so casually handled.

Instances have also occurred where plant engineering was being told by maintenance what type of nonmaintenance would or would not be done and by whom:

> In another case exhibiting poor control, a maintenance superintendent consistently stated that his workers could "do it for less" than the contractors identified by plant engineering. They rarely did, and it became a matter of serious dispute between maintenance and plant engineering. As costs escalated and maintenance quality fell, the plant

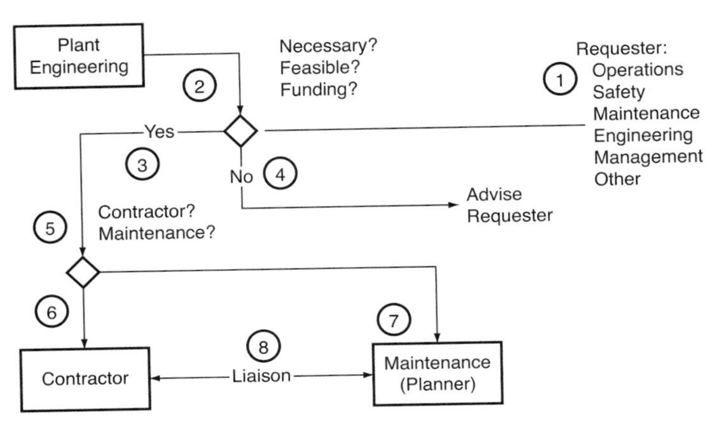

1. Requesters of new project work include operations, safety, maintenance, engineering, management, and others.
2. As these new projects are submitted, plant engineering should assess their necessity and feasibility, and verify correct funding.
3. If this assessment is positive, then the project would be considered for subsequent assignment to either a contractor or to maintenance.
4. If the new project does not meet approval criteria, the requester is advised. Approved projects should weigh how they are to be done by either a contractor or by maintenance.
5. For maintenance, their current work load and ability to complete the project within the required time frame is critical.
6. When work is assigned to a contractor, purchasing, acting under plant engineering guidelines, should select contractors according to strict quality performance criteria. In addition, contractors should meet commissioning criteria for turning equipment over to the plant for continuing operations and maintenance.
7. If the project will be done by maintenance, work should be planned to ensure its effective execution.
8. Finally, when work is done by a contractor, there should be a liaison with maintenance, who will continue to maintain the equipment after it is placed in service.

FIGURE 14-1 Flow of new project work

manager was forced to step in and establish specific policies and criteria for the conduct of nonmaintenance work.

HOW MUCH PROJECT WORK

Maintenance departments are often called upon to perform nonmaintenance work. Although the same craft skills are used, limits should be observed to ensure that this work does not utilize maintenance resources intended for regular maintenance work. Nonmaintenance work must not be allowed to interfere with the conduct of the basic maintenance program. Maintenance personnel are often perceived as equally competent in performing maintenance, as they might be in construction or modification. They are simply expected to get the work done. Although crafts are considered to be interchangeable, construction craftsmen often lack the diagnostic skills that distinguish them from maintenance craftsmen. A construction-type may relish "building a stairway to nowhere" while the maintenance-type wants to get on with restoring existing equipment. Therefore, a plant should start developing ground rules for handling nonmaintenance by first recalling the definition of maintenance:

> The repair and upkeep of *existing* equipment, facilities, buildings, or areas in accordance with *current design specifications* to keep them in a safe, effective condition while meeting their intended purposes.

Existing equipment, buildings, and facilities suggests that maintenance cannot begin until the new equipment, buildings, and facilities are created. *Current design specifications* suggests that modifications are not maintenance.

> Contractors often perform modifications or install new equipment. But maintenance must be involved so they can properly maintain the new facilities and equipment after the work has been completed. Maintenance must understand how the work was done and should expect certain minimum documentation by contractors, such as critical spare parts lists, wiring diagrams, maintenance instructions, and so forth.

WORK LOAD

Boundaries must be defined as to how much nonmaintenance work the maintenance department can perform and still be able to carry out basic maintenance. This requires the plant to know how much labor is required for maintenance. If the amount of labor required for maintenance is critical, nonmaintenance work should only be done during periods of low-maintenance work demand.

Plants should also scrutinize potential projects against other criteria such as

- Whether it is necessary. Some work may be proposed only to find that the equipment supports a product line that will be discontinued.
- The feasibility of the work. Will a modification of one unit adversely affect six others?
- Funding. Expensing nonmaintenance work will ensure maintenance budget overruns.

PLANNING

All nonmaintenance work, once assigned to maintenance, should be subjected to the same planning steps as maintenance. The scope and nature of the work should be evaluated. It should be awarded a priority so that it can compete fairly with other major tasks for the allocation of labor. During the weekly maintenance–operations scheduling meeting, nonmaintenance jobs should be considered on their merits. Maintenance supervisors should be precluded from accepting nonmaintenance work unless it has been approved in advance.

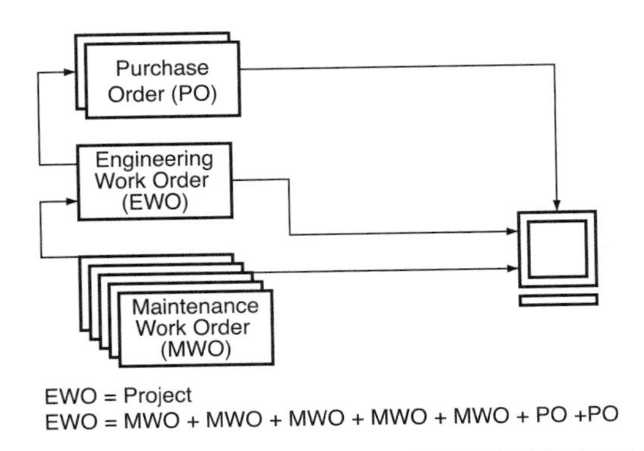

FIGURE 14-2 A single engineering work order (EWO) controls a major project. In turn, such a project might be carried out by maintenance, a contractor, or both. In this instance, one EWO controls a project carried out with five maintenance work orders (MWOs) and two purchase orders (POs) used by contractors.

Ground Rules

Preliminary ground rules for the conduct of nonmaintenance projects suggest that they

- Should be examined to ensure they are necessary and feasible;
- Must meet an approval and funding criteria;
- Should be planned and properly engineered; and
- On scheduling, should have resources allocated by priority.

PROJECT CONTROL

Often overlooked but necessary is the adequacy of the work order system for the control of nonmaintenance work. Because work may be done by either a contractor or the maintenance department, the work order system must provide a means for adequate control of this work. Usually this is accomplished with an engineering work order (EWO). It should complement the maintenance work order (MWO) and be compatible with the use of the purchase order (PO) to arrange for and control work performed by a contractor. See Figure 14-2.

Generally, the contractor's work is authorized by a PO, whereas an MWO is used by maintenance. Both must tie in to the accounting system. When resource use is reported, the contractor uses an invoice referring to the PO while maintenance uses a time card and a stock issue document. Typically, maintenance crews should

- Have adequate job instructions to ensure they can carry out the work properly. Don't make them improvise.

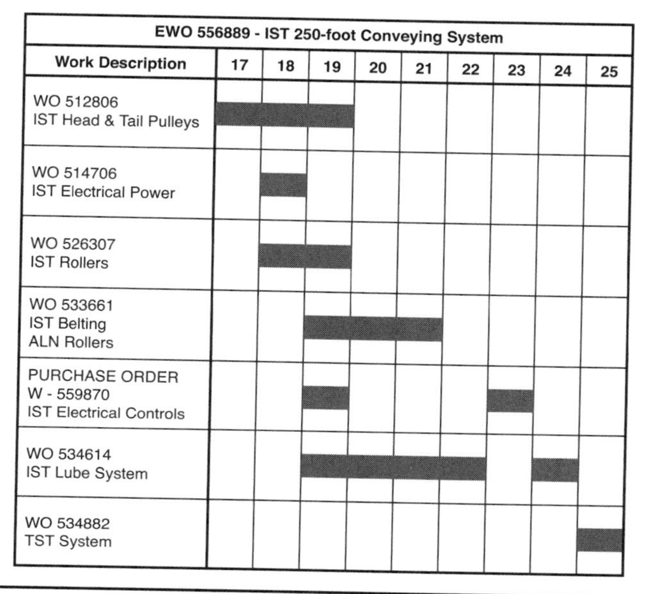

FIGURE 14-3 In this project to install (IST) a 250-foot conveying system using EWO 556889, the project engineer during week 21 can note that the work to install tail pulley (WO 512806), install electrical power (WO 514706), and install rollers (WO 526307) has been completed. He also notes that the contractor work to install electrical controls (Purchase Order W 559870) has been completed. He observes further that the work to install belting and align (ALN) rollers (WO 533661) is two-thirds completed and the work to install lube system (WO 534614) is half completed (2 weeks of the scheduled 4 weeks). Testing (TST) the system is completed in week 25 (WO 534882). As each work order accumulates labor and material costs, the project engineer can also observe the accumulated costs against each part of the project as well as the entire project.

- Get properly coordinated on-site support (e.g., materials delivery) or the use of supporting equipment (e.g., trucks for transporting people and materials) to ensure the project stays on schedule.

As the project is being conducted, information must be provided on the cost, project status, and performance of the work. See Figure 14-3.

At completion of the work, installations by either a contractor or maintenance personnel should be subjected to commissioning criteria so that maintainability is ensured.

CONTROLLING NONMAINTENANCE WORK

Although construction, equipment modification, installation, and relocation are not considered maintenance, many maintenance departments perform this work, sometimes to the detriment of basic maintenance:

- Excessive nonmaintenance work may cause maintenance to neglect basic maintenance. Thus, management should establish its limits.
- Maintenance is an operating expense; nonmaintenance is capitalized. Maintenance budgets should correctly reflect this difference.
- The work order system must control nonmaintenance work, but the MWO is inappropriate, especially with contractors.

These problems can be solved with good information, such as the following:

- Information on labor utilization allows management to determine whether nonmaintenance work is excessive.
- Maintenance can segregate nonmaintenance work and avoid expensing work that should be capitalized.
- A separate work order element for nonmaintenance work precludes misuse of the MWO.

The EWO is the means by which nonmaintenance (engineering) project work is controlled after it is determined to be necessary, feasible, properly funded, and then approved. It is not unusual for maintenance to utilize contractors for this work and even to work with them on the job. Therefore, a sufficiently flexible procedure is necessary to manage nonmaintenance work, especially if it is done by contractors. Nonmaintenance work like construction requires a feasibility determination before it can be funded and considered along with regular maintenance work.

The use of an EWO permits the project engineer to manage all aspects of the project through a single EWO. Details of progress and cost accumulation are provided by reviewing individual maintenance work order status reports, the contractor's invoices, and on-site progress. The project engineer would continue to monitor progress with frequent job-site visits, during which time the detailed job plan would be consulted. With the availability of good information on nonmaintenance projects, the problems of misuse of labor, incorrect funding, and project control can be avoided and replaced with effective project management.

PLANT ENGINEERING

Projects are temporary activities that have a definite beginning and end, as distinguished from ongoing maintenance work. Projects represent a temporary increase in the normal work load. Therefore, a maintenance organization asked to perform project work should carefully anticipate when their participation is needed, always weighing this temporary need against their full-time maintenance requirements. Typically, engineering will develop and complete each project in four phases (Figure 14-4). Maintenance involvement with plant engineering should parallel each project phase immediately after it has been determined that maintenance will perform the work. Of special importance is

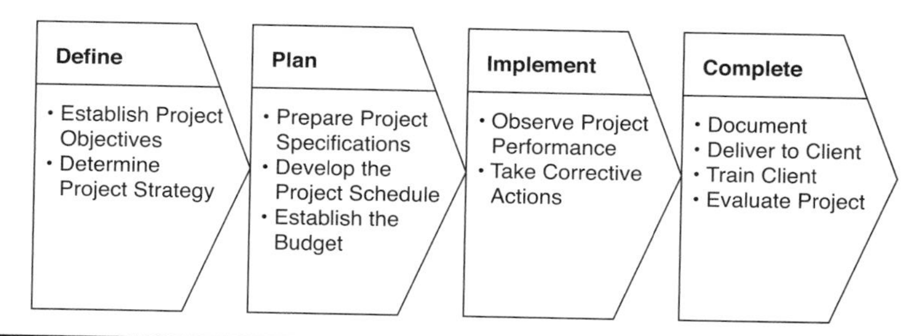

FIGURE 14-4 The four project phases

the documentation of the operating and maintenance details when the project is turned over to the plant (client) in the fourth phase.

Plant engineering should clearly define the project to include the criteria for determining successful completion of the project. Contingencies should be built into the project in anticipation of changes once the project is under way. Such changes should be documented and provision made to adjust both the project schedule and budget. Each project should result in equipment installation or modification that performs as expected. Projects should be completed on schedule and stay within cost limits. Thus, the project plan should define work quality in the specifications, time constraints imposed by the schedule, and budgeted costs.

When any project is carried out, a temporary team is organized under a project manager. Individuals are assigned duties and responsibilities and then trained. Specific policies and procedures clarify how the team is to function during the project. When work on the project begins, the project manager coordinates the tasks of different groups such as shops, maintenance, engineering, purchasing, and transportation. The project manager must monitor the progress of the project and measure it against schedules and budgets. As deviations occur, information on project status must be good enough to allow immediate corrective actions to be taken. As the project moves forward, project managers will provide feedback to team members and resolve problems on materials, supplies, and services. When the project is completed, documentation to include writing operations manuals, training of personnel on the use of the new equipment, reassigning of project personnel, and disposing of surplus equipment materials or supplies are steps that maintenance personnel must be aware of and may be called upon to perform.

Evaluating the project is the final step. This often includes an audit along with a project report and management review. In each instance, maintenance personnel will play a part. Because other projects will follow, it is especially important that maintenance provide direct input to ensure that future projects are made more efficient based on current project experiences.

UTILIZING CONTRACTORS

The most common use of contractors in the mining industry is fleet maintenance. Typically, an organization like Caterpillar provides an entire fleet of equipment and contracts to maintain them to a certain level of availability. In most other instances, contractors are used on capital projects like a plant expansion. Contractors who carry out the total maintenance operation are less likely to be found in mining as in oil and gas production, for example. The use of a contractor depends on how well the contractor can fit into the desired maintenance strategy. A contractor introduces a third party into an existing company team. Therefore, an assessment of whether a contractor can add "team potential" to the existing organization weighs as much as their ability to accomplish the required work. In general, if the contractor under consideration has personnel who can adapt and fit into the plant team environment, such a contractor could be successful. Therefore, guidelines for considering the use of the "right" contractor are appropriate.

A plant with a widely fluctuating annual work load is a good candidate for contracting maintenance, capital work, and plant services. Thus, most plants that have seasonal production requirements are candidates. For example,

> The sugar beet industry can only operate after the crop has been harvested and is ready for processing. During the growing season, plant employees are refurbishing selected equipment in anticipation of processing the next crop. During this period, no production personnel are required, and only a portion of the maintenance employees are needed. Often, unnecessary work is done to keep people busy.

Generally, there are three subelements of contractor use to be considered. Will the contractor perform:

- Only maintenance work?
- Maintenance plus capital nonmaintenance work such as equipment modifications and installation?
- Plant services such as sanitation or buildings and grounds work?

Considering each possibility, there are then different ways that the work can be managed and carried out. If the contractor will perform only maintenance work, the contractor work force might consist of

- Only supervision and workers with the host plant providing management and planning, or
- The total maintenance requirement of management, planning, supervision, and performing the work.

Similarly, if contracting maintenance and capital work were being considered, the only difference would be the role of the plant engineer in specifying and managing capital work performed. If only services such as grass cutting are involved, the contractor merely shows up at the proper times and a mechanism is arranged to verify the quality of the work and to oversee administrative arrangements.

Objective

Next, the objective of the contract work should be established:

- Will the contractor supplement the work force?
- Will the contractor be the work force?
- Will the contractor be used for specific projects such as annual plant turnarounds, seasonal work like grass cutting, or full-time work such as custodial services?

Policies

When it has been established that a contractor could benefit the plant and their potential use is established, a policy regarding the use of contractors should be established. This policy should prescribe the boundaries of the contractor's work and their working relationship with plant employees. If the use of the contractor will displace existing employees, it should be coordinated to avoid future conflict in labor relationships. Policies should be widely publicized to ensure that they are understood by all plant employees.

Ongoing Programs

The maintenance program and the methods for carrying out capital work should be explained to the contractor candidates. In doing this, the discussion should include how these matters will be accomplished:

- How will the overall task be managed?
- Who will do the planning?
- To what degree will the contractor provide resources?
- Will the contractor provide their own information system if they are to control the work?
- How will materials be supplied to the contractor for maintenance and capital work?
- Will supporting transportation, rigging, and lifting equipment be provided by the contractor or the plant, and who will operate them?
- Will this equipment be maintained by the contractor or the plant?

In addition, issues like quality control or continuous improvement programs must be carried out by the contractor to the same level of compliance as the plant. If programs like reliability centered maintenance are to be used, they will require the application of effective condition-monitoring technology. The contractor's personnel must be prepared to use such technology. They must also provide training to ensure that future technology applications can be effectively utilized by their personnel.

Performance

How will the contractor's efforts be assessed? What measures will be applied to judge the satisfactory accomplishment of individual jobs and the management of the overall contractor effort? Will the plant:

- Judge the contractor on the overall reliability and availability of equipment?
- Measure accomplishments like weekly preventive maintenance compliance or completion of a shutdown on time?

How will these measurements affect the reporting required by the contractor and the ability of the plant's information system to present performance to local management? What specialized staff personnel might the plant have to provide to oversee the contractor's efforts? The type of work being done by the contractor and the working relationship between the plant and the contractor will influence the degree of control and the means by which it is accomplished. In the instance of a large plant where the contractor might assume responsibility for maintenance and capital work, a gradual transition ensures continuous control as the contractor is phased in. In addition, the degree to which the contractor can successfully absorb responsibilities can be assessed as the transition takes place.

This transition could be especially helpful in the implementation of any new strategy for a contractor more accustomed to the traditional control of plant maintenance and capital work. Consider the following phasing-in sequence:

1. The contractor provides only supervision and workers.
2. The contractor replaces plant maintenance planning personnel.
3. Engineering staff are provided by the contractor.
4. The plant information system supports the effort.
5. The contractor's information system links with the plant system.
6. The contractor uses the plant warehouse.
7. The contractor takes over the plant warehouse.

Flexibility

The most prominent reason for the use of a contractor is the desired flexibility provided by the contractor's work force. Often, the plant's year-round fixed work force is seen as a financial liability if the product demand is uneven. Therefore, seasonal demands, shutdowns due on short notice, or the need to cancel capital work as conditions change require that the contractor's work force be flexible with the number of people it can provide or withhold. Plants, on the other hand, wish to avoid the labor cost penalty with a fixed work force when they have little work to do.

Auditing

Assuming that the contractor's work force is in place and performing significant maintenance and capital work, an annual auditing procedure should be applied. This is in addition to the periodic performance checks applied. The audit should embrace every aspect that touches the work the contractor does. For example, if the contractor now runs the stock room, criteria should be applied to assess stock outages and quality of service. In addition, personnel should be interviewed, especially in operations, to ascertain whether they feel, subjectively, that their equipment is now more reliable and productive. The audit should be followed up with recommendations for improving contractor services and a reexamination of areas needing better performance.

Payment

Contractor payment might include incentives to accomplish work more efficiently or to add value to the quality of work by improving work methods. For example,

- The cost of labor to install each dollar of material has been reduced.
- The contractor's training program has allowed them to use multiskilled personnel, resulting in a smaller contractor work force.
- Contractor personnel have mastered needed equipment-monitoring techniques resulting in greater equipment availability.

SUMMARY

Plant maintenance departments are often called on to perform nonmaintenance work. If the plant has a dedicated construction staff, the process can be carried on routinely. However, if the nonmaintenance work will be done along with regular maintenance work, steps must be taken to ensure that it is done with proper consideration so that it does not interfere with accomplishing the basic maintenance program. In the mining industry, the potential conflict between excessive nonmaintenance work and ensuring that basic maintenance is being done is often resolved by using contractors on a short-term basis. When all other considerations for contractor use have been applied, the best candidate of those identified can be chosen in consideration of the needs of the plant. If however, a contractor will perform direct or indirect maintenance tasks, one must be aware of evidence that they play a team role and verify that they are adapting to any new strategies and are providing reliable cost and performance information.

CHAPTER 15

Benchmarking

INNOVATIVE ADAPTATION

Benchmarking is the systematic process of searching for best practices, innovative ideas, and highly effective procedures that lead to superior performance (Bogan and English 1994). Another view has it that benchmarking is like watching a magic show from the wings in which all of the magician's tricks are revealed. Traditionally, benchmarking is a standard by which others may be measured. The idea of learning from others' good ideas or benefiting from their successful experiences as well as avoiding their mistakes is a philosophy of long standing. Thomas Edison, for instance, remarked, "Keep on the lookout for novel and interesting ideas that others have used successfully. Your idea has to be original only in its adaptation to the problem you're currently working on." Commenting on the avoidance of unnecessary problems, Prince Otto Von Bismark said, "I prefer to profit by other's mistakes, and avoid the price of my own" A perennial question: "What is it that organizations do that get results so much better than ours?" (Juran and Gryna 1988). There is much to be said for the benefits of learning from the experiences of others and adapting them to one's own situation. This can save time, avoid "trial and error," or speed up the process of change. But one must make certain that what is benchmarked has value to one's improvement needs.

BENEFITS OF BENCHMARKING

There are three major benefits derived from benchmarking (Bogan and English 1994):

- Cultural change—Benchmarking causes people to realize that others do things differently and they are motivated to find out why. In the process they find out that different may also be better.
- Performance improvement—Gaps in performance are identified and personnel are motivated to develop means to close the gaps.

- Human resources—People see the gaps between what they are doing and what others have accomplished and clamor for training to help them close the gaps (Ross 1995).

USING BENCHMARKING

Benchmarking is a comparison of pertinent practices, not a contrast of performance. Thus, the "model" an organization aspires to must be reasonably similar. Therefore, one must be aware of differences between one organization and the one being benchmarked, but at the same time, be aware that benchmarking with dissimilar organizations can yield ideas of value as well. For example, a process for stock control in an unmanned warehouse of a mail-order house had direct applicability to a food processing plant maintaining forklifts in five warehouse locations. But because the main concern is improving maintenance, the starting point should be other maintenance organizations. The following questions should be asked:

1. Do they have a similar industrial maintenance situation with, for example, a need to perform nonmaintenance work along with basic maintenance while using the same personnel for both?
2. Has their new team organization or the use of business units been successful, and if so, why?
3. Has their program been well-documented and is there evidence that all personnel understand it and support it effectively?
4. What value do they place on good information, and what are the prospects that their information is timely, complete, and accurate?
5. What specific ways do they control their work, considering organization, information, program details, leadership, motivation, and employee empowerment?
6. Have they successfully implemented new maintenance strategies, and if so, how did they accomplish this?

TYPES OF BENCHMARKING

There are three types of benchmarking: process, performance, and strategic (Bogan and English 1994).

Process benchmarking deals with maintenance processes and the methods people use to carry them out effectively. Typically, process benchmarking inquiries might examine how quickly and effectively reliability centered maintenance was implemented.

Performance benchmarking aims at how a whole operation brought together their management, production, engineering, and maintenance personnel in a successful overall operation. Generally, performance benchmarking

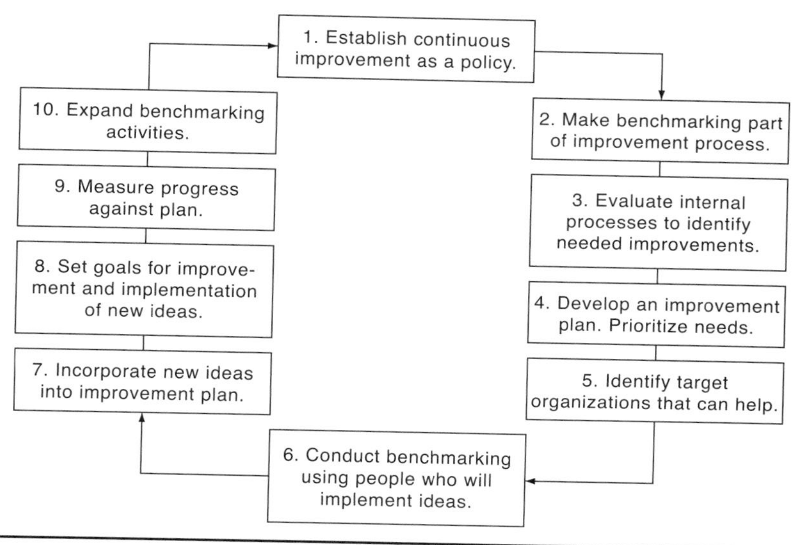

FIGURE 15-1 Benchmarking steps

would study how the organization successfully implemented total productive maintenance throughout the plant.

Strategic benchmarking transcends any particular industry. It explores how a culture or nation carries out a successful industrial campaign to significantly impact their world standing. Strategic benchmarking might, for example, inquire how Japan became a world leader in electronics even though the United States led the industry until the early 1980s.

BENCHMARKING STEPS

Benchmarking should be part of an overall improvement process. It should be thought of as an integral action of the improvement activity, not a one-time "look see" to satisfy those curious about what others are doing. Specific steps, such as those outlined in Figure 15-1 and described in the following paragraphs, should be established to ensure that the benchmarking process is worthwhile and that is satisfies specific, needed improvement requirements. This will establish immediate legitimacy and get the process off to a good, useful start.

Step 1. Establish continuous improvement as a policy. Management should include any useful method that will contribute to improvements. Benchmarking is such an activity. The more formally it is approached, the more effective the results will be.

Step 2. Identify benchmarking as an integral part of continuous improvement. With official sanction, benchmarking becomes part of the planning to achieve improvements in a serious way.

Step 3. Conduct an internal evaluation to establish current performance level and improvement needs. The first step of any improvement action is to establish the "as is" performance status of the organization and identify its improvement needs. This step is critical to the benchmarking process. Without it, one does not know where to begin, much less what will help and what won't.

Step 4. Develop an improvement plan; then classify and prioritize improvement needs. The benchmarking exercise can be a big waste of time if there is no overall plan of action to guide benchmarking personnel. They will have little idea of what to look for and where to find it. Make a plan for improvement based on the evaluation results. Then fit into that plan into the procedures for benchmarking and the specific things that are being sought through benchmarking. This will guide those doing the benchmarking in fitting the results into a bigger improvement requirement.

Step 5. Identify "target" organizations that can help. Look in your own organization. Then look in your own industry. Finally, look for the best. But don't overlook other organizations that may be doing other things. After all, it is what people do and how they do it successfully that is the central value of benchmarking.

Step 6. Conduct benchmarking using personnel who will ultimately implement improvements, including the ideas derived from benchmarking. Direct participation by those who will implement the new ideas means that they start to formulate actions on how the ideas can be used as they have heard them. This ensures enthusiasm and interest.

Step 7. Incorporate the new ideas developed from benchmarking into the improvement plan. New ideas should be related to the initial needs established at the time of the internal evaluation. In this way, more impetus is added to the initial need. Benchmarkers know they have potential, immediate solutions to improvement requirements.

Step 8. Set goals for improvement and implementation of new ideas developed from benchmarking. The combination of the action plan from the early evaluation plus the new ideas from benchmarking will suggest a new priority-setting action to update the improvement plan. This is the time to add emphasis to the improvements with a timetable, goals, and targets. It gives managers an opportunity to reinforce the initial plan and acknowledge the work of the benchmarking team.

Step 9. Measure progress against goals and the overall improvement plan. With the revised plan of action and renewed enthusiasm for improvement, establish goals to keep everyone apprised of progress. More importantly, provide a means of encouraging forward progress with regular "report cards."

Step 10. Expand benchmarking to other industries or activities. Cast the net wider now that benchmarking is established and considered an integral, important aspect of continuous improvement. Empower employees to go forward on their own to seek new ways to improve.

BENCHMARKING QUESTIONS

After a decision has been made to benchmark, the "model" or target organization should be carefully identified. Ensure that what they are doing fits your needs. Here are some questions that should be raised: Do they have the same type of maintenance organization? Is their maintenance program the same as yours? They may consider modification as maintenance while you do not. They may not define how they conduct their work and may even be organized differently to carry it out. They might, for example, consider that overhauls are preventive maintenance while you do not. These are the types of close comparisons that must be made as the benchmarking task force gets organized.

COMPARABLE PERFORMANCE MEASUREMENTS

Typical maintenance performance indices often include the percentage of work planned or compliance with the preventive maintenance program. Although these are reasonable, they will be very misleading if the target organization has different criteria for determining what work is planned and how it defines preventive maintenance. Thus, sufficient detail is necessary to determine whether there is reasonable compatibility between the methods by which activities are controlled. The following key points should be considered when comparing a target organization for benchmarking purposes.

- **Key personnel.** Do the target organization's key people perform the same job functions that people with comparable job titles do in your organization? Their maintenance engineer may be a project manager for capital work while yours is concerned with actions to increase equipment reliability. Their planner may simply schedule routine actions while yours plans major jobs exclusively. Identify comparable actions with less attention to job titles.
- **Use of operators.** Are the roles of the target organization's operators comparable with yours? Their operators may not do any maintenance, yet yours seem as if they are part of the maintenance department. But their operators produce well and their maintenance costs are enviable. Find out why.
- **Nonmaintenance.** Does the target organization carry out, control, and fund nonmaintenance the same way as it is done in your organization? If their capital work, like modifications, is expensed while your organization capitalizes it, cost–performance comparisons are likely to be very confusing. Look for common ground before benchmarking.
- **Value of information.** Is information important to the target organization? If cost is said to be important and maintenance craftsmen do not report labor detail, then someone else is guessing at it. If so, the labor portion of their cost information is incomplete and probably inaccurate. Thus, the quality of other information may be suspect, and any

claimed achievements based on information will require verification before you accept their validity.

- **Overall program.** Can the target organization explain their overall program? Start by asking that they explain their overall maintenance program in detail. If the explanation is clear and well-documented, you can learn from them. If it is not, check any program elements very carefully before you accept their claimed accomplishments. What you hear may be the maintenance manager's fantasy.
- **Adequate evaluation.** Will "perceived" gains stand a real evaluation? Ask whether their total target organization has been evaluated and what the results were. If there has been no evaluation, their accomplishments may be unfounded speculation.
- **Phony accomplishments.** Are accomplishments real, or are they based on the boss's "claims"? Find out what documentation they have to support the idea that their accomplishments are worth transferring to your organization. Delve deeper into the organization where you may find that the real situation is less "rosy" than the boss says it is.
- **Subtle but meaningful differences.** If your organization performs construction work using maintenance personnel, and the target organization has a dedicated construction crew, you must realize that their construction is significantly different from yours, with little in common, so benchmarking will find little to look at.
- **Critical points.** Pay less attention to the detail of what the target organization is doing and more to the way they have organized and utilized their personnel. If, for example, worker productivity is proven to be outstanding, find out the organizational and leadership steps they took to improve it. Pay less attention to the claim that productivity is due to planning 60% of their work. If equipment availability is high in a similar organization, examine what they did with their people to get it that way.

  Keep in mind that improvement comes about largely through education, motivation, and leadership. Therefore, pay primary attention to how the benchmarked organization used their people more effectively. Prepare for the benchmarking visit by doing your homework. Know what you are looking for. During the benchmarking process, imagine that you are carrying out an evaluation of the target organization with specific performance standards in mind. Take charge of the benchmarking with your agenda. Avoid giving the target organization the opportunity to tell stories. Insist on proof and good documentation before acting on activities that only appear to be beneficial. Beware of the plant that seems to do nothing but host benchmarking visits. Most likely, the manager has a good story to tell but little substance. His or her goal may be to achieve vice presidency based on publicity.

BENCHMARKING CHECKPOINTS

Potential organizations with whom benchmarking could benefit an organization should exhibit specific characteristics. All characteristics should contribute to their success and increase a company's desire to explore more closely how they achieved that success. The following questions should be explored to determine how they were able to be successful.

- Why did their plant manager spell out their production strategy so carefully and go to such trouble to integrate maintenance into it?
- How did they organize and what are the special ways that they operate?
- What steps did they take to establish quality work standards, and how have they sustained them?
- How do they measure the reliability of their equipment, and what steps did they take to gain reliability and sustain it?
- How did they make their overall maintenance program effective?
- What steps did they take to improve productivity, and how do they ensure continuing productivity?
- What did they do to get the various staff departments to be supportive and provide such a high level of service?
- What special relationships did they establish with warehousing and purchasing to ensure that the right materials for maintenance are consistently available?
- What steps did they take to identify the right technology to ensure reliable equipment with the application of the best maintenance techniques?
- How are they controlling maintenance work to yield such an effective level of cost control?
- How did they successfully integrate ongoing performance measures with their performance auditing to ensure they are able to identify and correct problems quickly?

LEVELS OF BENCHMARKING

Starting from the most basic effort to reduce one's mistakes, benchmarking progresses upward to checking out the "world class" acts. Here is the progression:

1. **Learn from mistakes.** If something has gone wrong, find out why and develop procedures or educate personnel to ensure it does not happen again.
2. **Learn from successes.** When something goes well, make certain that the circumstances behind the successes are incorporated into ongoing practices.
3. **Borrow any good ideas.** Check out the good ideas from anyone, and, if useful, shamelessly borrow them and apply them to help solve your problems.
4. **Apply internal best practices.** Look around your own organization as an additional source of good practices. Practice "management while walking around." You will see many things you did not realize were going on. Many will have real value.
5. **Compare with a leader in your industry.** Identify the best in your industry and work toward an opportunity to visit them.
6. **Look beyond your industry.** With people actions, other organizations doing significantly different things can be a useful source of worthwhile ideas. An electronics manufacturer can gain significantly by borrowing an order-entry process from a chain store, for example.
7. **Assess the world-class level.** Benchmarking really establishes what your accomplishments actually are. Therefore, a world-class organization becomes the standard to meet.

SUMMARY

Benchmarking can be useful as a guideline but misleading as a performance measure. Successful benchmarking must be based on actual improvements needed as the result of an evaluation. Otherwise, there is only conjecture as to what could be gained from benchmarking. Make benchmarking part of the improvement strategy to give it credence, and ensure that it contributes useful ideas to yield actual improvements.

REFERENCES

Bogan, C.E., and English, M.J. 1994. *Benchmarking for Best Practices: Winning Through Innovative Adaptation*. Chicago: R.R. Donnelley & Sons. pp. 1–2.

Juran, J.M., and Gryna, F.M. 1988. *Juran's Quality Control Handbook*. New York: McGraw-Hill. pp. 95–96.

Ross, J.E. 1995. *Total Quality Management*, 2nd ed. Delray Beach, FL: St. Lucie Press. pp. 239–240.

CHAPTER 16

Material Control

MATERIAL IDENTIFICATION AND PROCUREMENT

Maintenance organizations require that the right materials, in the proper quantity, be available at the right time. Material control departments (purchasing and warehousing) provide procedures to identify stock and purchased materials in proper quantities and make them available to maintenance as requested. In turn, maintenance must determine what it needs and specify the quantity and the time they are required. Material control procedures should specify material control responsibilities for planners, supervisors, and crew members. Generally, planners obtain the materials required for planned work. The supervisor and his or her crew members obtain materials for all other types of work.

REPLACEMENT FORECASTING

The maintenance program will include material requirements. One of the most useful elements that can contribute to the more effective control of materials is major component replacement forecasting. By identifying the future time when major components might be replaced, an advance need for materials is announced, enabling material control to respond more effectively (Figure 16-1). Standard bills of materials are used to help expedite planning. But these bills of materials also make procurement faster, more efficient, and reliable. Both maintenance and material control benefit. When the forecast is shared with purchasing, for example, they can anticipate requirements for the future and get ready. Similarly, the warehouse can prepackage kits of materials or tools for major jobs. (Forecasting is discussed in more detail in Chapter 9.)

MATERIAL CONTROL INFORMATION

Linkage between the work order system and the inventory control program permit maintenance personnel to identify and order stock materials more easily. See Figure 16-1.

CC	TP	UNIT	CP	STOCKED PARTS FOR COMPONENT	
00	01	000	05	141576	Hydraulic hose 1/2"
				141588	Fitting 1/2"
				141592	Adapter 1/2"
				141598	Sleeve clamp assy
				141599	Sleeve clamp bracket
				141602	...

FIGURE 16-1 In this illustration of a typical computer screen, the user selects the equipment type (TP) and the component (CP). The computer responds with all of the stocked parts to fit the component for the type of equipment and component specified. To order selected materials, the user highlights parts needed, adds the quantity of each plus the work order number, and transmits the order to the warehouse. Later, the user picks up the parts at the warehouse or arranges to have them delivered to the work site.

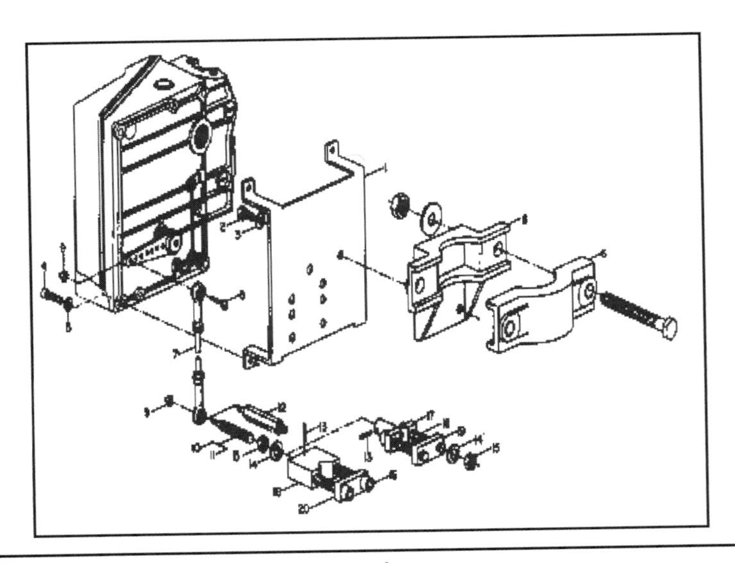

FIGURE 16-2 Computer graphics example

The availability of computer graphics helps as well. Both maintenance and material control personnel are able to quickly identify parts as well as see exactly how a component is reassembled. Moreover, these graphics can be printed and taken back to the job site by craftsmen for reference as they re assemble the components or units. See Figure 16-2.

PURCHASING INFORMATION

Few organizations want to incur the cost of stocking materials if they can purchase them as needed. The cost of stock materials versus purchased materials is changing. Generally, the advent of better information has made a difference. For example, the need for major assemblies can be better anticipated. Thus, orders can be placed to suppliers with assurance that on-time delivery will result. Inventory control systems of several operations can communicate. Thereby, stocked materials not available at one site can be "borrowed" from another site. Although maintenance should avoid the role of the "professional" purchasing agent, they often have a requirement to identify materials that must be purchased. If there is not time to plan the anticipated work, maintenance supervisors may have to make inquiries directly with suppliers. However, only the purchasing agent can commit the funds. Thereafter, maintenance personnel should have access to purchasing information, allowing them to obtain the status of any purchase order. This is especially true of the planner, who handles most purchased material needs.

PARTS MANUFACTURING

When parts or fabrication are required for planned jobs, maintenance field planners coordinate with shop planners to get the work done. However, when stocked parts are to be manufactured, it is the warehouse supervisor or the purchasing agent who works with the shop planner. Therefore, a close working relationship and compatible systems are desirable. Generally, parts being manufactured for stock require a determination of minimum quantities because the machining setup costs can be significant. Similarly, maintenance should not misuse the capabilities of their shops to manufacture, for example, a part for $600 that could have been purchased for $12. Remote mining locations far removed from cities with major parts distribution centers must have a comprehensive, effective procedure for duplicating spare parts. Similarly, standardization of equipment can often avoid the need for stocking a multitude of different spare parts.

COMMERCIAL SHOPS

Commercially manufactured stock parts are the domain of the purchasing agent acting on needs expressed by the warehouse. However, when major components are rebuilt by a commercial shop, there is a need for maintenance involvement. Generally, components to be rebuilt are delivered to a classification point by maintenance. Once classified, components are sent to the authorized vendor's shop. These transactions are controlled with a purchase order. Vendors deliver the rebuilt components to the warehouse where they are placed back in stock. Once issued, maintenance should be able to track component performance by serial number in the repair history.

ON-SITE PARTS DELIVERY

The ability of the warehouse to deliver stock or purchased materials to the work site contributes to maintenance productivity and performance, and it should be encouraged.

STANDARD BILLS OF MATERIALS

A link exists between the maintenance work order, stock issued, and materials purchased for specific planned jobs. If any of these jobs are repeated, a standard bill of materials can be developed. Then, upon determining that the job will be due again, the same materials can be ordered in advance.

COMPONENT REBUILDING

A major service of warehousing and purchasing is their support in the classification, rebuilding, and restocking of components for future use. See Figure 16-3.

EXPECTATIONS FROM MATERIAL CONTROL

Maintenance expects the material control departments to

- Ensure that material control procedures are easy to follow and effective,
- Provide easy access to critical spare parts lists,
- Make sure that needed items and quantities are properly stocked,

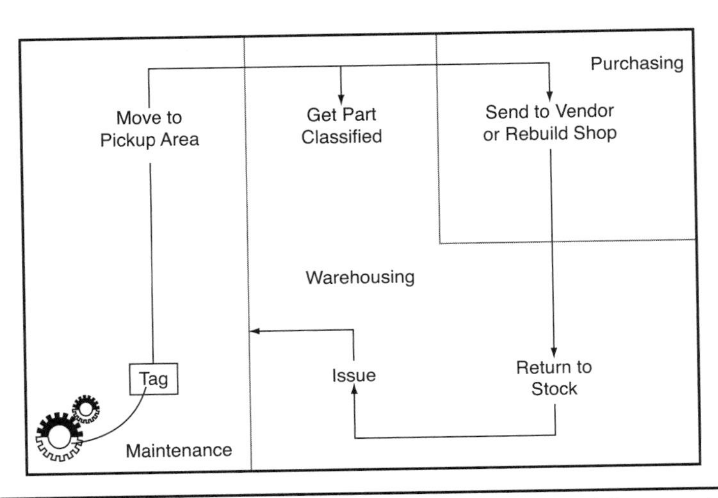

FIGURE 16-3 Tagged, damaged components are moved to a pickup area in the warehouse for classification. If rebuildable, they are sent to the vendor or a local shop for rebuilding. Once rebuilt, they are returned to stock for subsequent issue.

- Promptly replenish stock materials,
- Keep the stock room properly staffed with qualified personnel,
- Make stock withdrawal efficient,
- Provide easy-to-use stock return procedures,
- Ensure proper stock accountability,
- Make stock materials easy to identify,
- Provide for interchangeability of parts,
- Provide procedures for reserving stock parts,
- Help in developing standard bills of materials for repetitive jobs,
- Make issues on presentation of approved work orders,
- Summarize material costs by equipment and component,
- Provide material costs job by job,
- Provide for purchase order tracking,
- Provide drawings for remanufacture of parts, and
- Deliver materials to job sites.

Often craftsmen have had little material control experience requesting parts. Most are uncertain of procedures and need help and encouragement. They are most concerned with obtaining materials while jobs are in progress. They must identify and obtain materials quickly to avoid prolonging repair time. Even though the planner may have done a good job in getting most materials during planning, the actual work often reveals a need for additional, supplemental stock parts. Getting them quickly and easily is the craftsman's responsibility. He or she is also concerned with the availability of free issue items, the availability of large crew and power tools, the steel supply for field fabrication, and field support items such as ladders, rigging, slings, and cables.

Supervisors are interested in computer capabilities for parts identification as well as reserving parts. Then as daily assignments are made to crew members, they can be better assured that materials are ready. Emergency work often requires fast response in which maintenance and warehousing have established standard procedures in advance. Generally, the warehouse, rather than purchasing, should be the primary source of the supervisor's material needs. Planners, on the other hand, will deal equally with warehousing and purchasing. The planner's sources of materials include purchasing, warehousing, local shops, vendor's shops, and the use of contractors. Planners must be especially aware of the lead time required for shops and purchasing to obtain materials. Similarly, because the major planned jobs require crew tools, planners must provide the same lead-time considerations to tool room personnel. Planners' material needs are divided into two phases. In the first phase during planning, they identify and order purchased and stock materials plus shop work. In the second phase, after operations approves the schedule for the major jobs, additional stocked parts may be ordered, tools are reserved, and rigging or transportation services are arranged.

EXPECTATIONS OF MAINTENANCE

Material control personnel expect maintenance to follow the basic rules of accountability. Material control personnel also want to know what is going on. Thus, if maintenance has a forecast for the timing of major component replacements, they should share it with purchasing. Material control personnel should also be included in operations–maintenance scheduling meetings. They can contribute significantly to solving material problems when they become aware of them.

OPEN COMMUNICATIONS

Every opportunity to exchange ideas about improving material control effectiveness should be taken advantage of. This is where the essential coordination between the departments can be most effectively reinforced, as portrayed in the following situation:

In a large, open pit mining operation, open but covered storage of major, heavy assemblies was necessary to permit ready access by maintenance personnel at any time. Because the mine operated several pits and distances were great, these storage areas could not be staffed by warehouse personnel. After conferring on the problem of accountability, both departments agreed to try a bar-coding procedure that would record the departure of an assembly from a storage area. Although this procedure allowed accounting to be able to reorder the assembly, maintenance needed a way to record the repair history and associate the material cost of the assembly with the labor used to reinstall it. After collaborating again, a joint procedure was worked out to link the work order with the bar-coding procedure. Everyone could now meet all their management requirements more effectively.

EQUIPMENT REPLACEMENT

The decision to replace production equipment as it nears the end of its life cycle, based on considerable experience in operating and maintaining the equipment, is an economic and performance decision. Economics are involved in the identification and procurement of the best equipment for the capital outlay constraints involved. Performance considerations are important in selecting the best equipment for the job.

ECONOMIC CONSIDERATIONS

A whole host of economic-related factors must be weighed as the replacement decision is being considered. Although most of these factors are cost oriented, just as many are tied closely to the cost of delivering the required performance demanded of the new equipment. Therefore, as an organization goes about the task of replacing equipment, accounting considerations must be balanced against performance considerations. Field personnel must be an integral part of the decision-making process. A new piece of equipment might be obtained at a competitive price, but once in operation, if it proves difficult to operate and maintain, the savings at purchase can be quickly canceled with production losses. Economic and performance factors must be examined, weighed, and compared as replacement decisions are made (Table 16-1).

Definitions of Economic Considerations

Operating cost—The total cost of operating the unit of equipment. Mobile equipment operating costs would include consumables such as fuel, oil, lubricants, tires, and so forth, plus the wages and benefits of operators. Fixed equipment operating cost includes pro-rata power consumption and modifications required to meet safety and environmental standards or improve performance.

Maintenance cost—The total cost of labor and materials, including consumables like shop stock (welding rods), plus the use of mobile equipment (e.g., a mobile crane or transportation to and from work sites) to keep this equipment in a safe, effective operating condition.

TABLE 16-1 Economic and performance considerations for equipment

Economic Considerations	Performance Considerations
Operating cost	Ease of operation
Maintenance cost	Ease of maintenance
Depreciation	Safety needs
Cost of obsolescence	Environmental considerations
Acquisition cost	Status of replacement parts
Replacement cost	Technical support
Utilization of equipment	Overhaul vs. replacement
Impact of inflation	Productive capacity
Cash-discounting	Maintenance facilities
Financing	Operator training
Tax considerations	Maintenance training
Resale or trade-in value	New tools and procedures

Depreciation—The loss in value of the equipment over time.

Cost of obsolescence—The reduced market value of the equipment to another operation as the unit becomes older in terms of wear, dependability, inadequate productive capacity, and less desirable features, such as controls, ease of operation, and so on.

Acquisition cost—The current cost to obtain a suitable replacement unit.

Replacement cost—The overall cost of replacement, including removal of the old equipment or commissioning of new equipment in a ready-to-be-operated condition.

Utilization of equipment—Considerations of how the new equipment versus the old equipment will be utilized. Older equipment may have been used in a different productive capacity that is being phased out. For example, haulage roads in an open pit mine have become steeper as the ore body is mined. Thus, the new haulage trucks must be assisted with an electric-powered trolley to compensate for the additional fuel consumption of the steeper haulage roads. The new trucks will require configuration for the trolley assist.

Impact of inflation—Labor to operate and maintain the equipment as well as repair parts, fuel, and power will all be more expensive because of inflation when the new equipment is made operational when compared with current costs.

Cash-discounting—A discount in the transaction that may be allowed if the buyer can pay in cash, thus avoiding interest charges and accompanying administrative fees. The equipment seller or the organization to whom old equipment is being sold may find that a lesser amount paid at purchase or at sale is more desirable than the agreed-upon amount paid later or over time. Thus, an immediate payment could result in a cash discount.

Financing—The cost of borrowing the money to make the capital outlay for new equipment is a consideration. Although banks are a primary source, often equipment manufacturers can discount financing costs as a result of considerations such as warranties, involvement in installation or commissioning equipment, and even third-party arrangements with selected banks. In certain developing countries, World Bank funds may even be a factor.

Tax considerations—Plant expansion that brings new jobs to a region or local tax rebates to locate a plant or operation in a certain place could reduce state or local taxes to a plant or operation. Often, early compliance with environmental standards may allow a rebate of federal taxes.

Resale or trade-in value—The market value of the old equipment as determined by the perception of the prospective new user.

Performance Considerations

Ease of operation—How easily can operating personnel operate the new equipment? Field personnel should be given the opportunity to operate the new equipment being considered and talk with operators already using the equipment. Often, there may be some quirks that could prove difficult in the retraining of operators or the use of the equipment. Similarly, personnel could be enthusiastic about the new equipment in every respect and eagerly endorse the new equipment over its competitors.

Ease of maintenance—How easily will maintenance personnel be able to carry out repair and maintenance? For example, all components should be easily accessible and there should be clear, easy-to-follow maintenance procedures on the new equipment.

Safety needs—Can the equipment be operated safely, and are there devices and procedures in place that ensure safe operation? Operators, maintenance personnel, and trained safety specialists should examine the equipment for possible safety hazards that could occur in the work place if the new equipment were to be selected.

Environmental considerations—Will the equipment once placed in operation meet current environmental standards? Determinations must be made as to whether the new equipment will comply with environmental standards being imposed now and whether the equipment will need extensive modification in the near future to meet standards that will be imposed.

Status of replacement parts—Will replacement parts be easily available for either immediate use or with adequate advance notification to parts suppliers? The equipment that represents the recent massive capital outlay is a total loss if it is not producing because replacement parts are not available or an unreasonable period of time is required to get them to the plant site. Therefore, every possibility that this will not happen should be raised as the negotiations surrounding the new acquisition are carried out. Among the points to be considered are standardization of parts, local duplication or remanufacture if in a remote location, consignment stock considerations, guaranteed delivery in hours or days, and instructional manuals on parts installation.

Technical support—Is technical support readily available for unusual maintenance problems? New equipment typically brings new problems and questions. Thus, technical support—whether by telephone, fax, e-mail, or CD—must be considered. In addition, the availability of field technical personnel in the case of especially complex or dramatically different equipment is important.

Overhaul versus replacement—Will guidelines be provided to assist in a determination of whether the equipment should be overhauled or replaced? If the general utilization of the equipment will not be changed and the current features are considered adequate, can a few more years of utilization be secured by overhauling the existing unit rather than replacing it?

In this instance, the estimated cost of the overhaul must be carefully weighed against the general performance of the unit in terms of satisfactory productive capacity, reasonable repair, and cost history. Quality maintenance information must exist to be able to consider the overhaul-versus-replacement option.

Considering that the production equipment of the new century will have greater productive capacity and inherent reliability built into it, more than likely it will also be more complex. Maintenance people might be in for some real surprises in terms of skills required, new techniques to be learned, or the need to acquire special tools before they could expect to maintain the equipment properly. On the other hand, critical components might have easier access, be simpler to exchange, or be designed to last longer. Those maintenance people who do the actual repair should be asked to help evaluate the equipment under consideration.

Productive capacity—Will the new equipment meet the desired capacity of production requirements? Acquiring greater productive capacity with new equipment in a plant or operation may determine the future profitability of a company. Thus, the needed capacity for now as well as in the future must be carefully identified as candidate equipment is being identified.

Maintenance facilities—Will the existing maintenance facilities such as bay space or crane capacity support required maintenance of the new equipment? Operating and maintenance personnel must carefully advise on the capacities of facilities that will be required when the new equipment is placed in operation. Points to consider: Will new or bigger maintenance facilities be required to maintain the newly acquired mobile equipment? Is the lifting capacity of overhead cranes in the plant sufficient to move materials or replacement components to service and maintain the new equipment?

Operator training—Will operators have to be significantly retrained to operate the new equipment or can they simply be reoriented? How much of a retraining program will be required for operators to be able to operate the new equipment safely and effectively? Is the new equipment operation so different that different skill levels must be established and new labor agreements negotiated once the equipment is in operation?

Maintenance training—Can maintenance personnel simply apply existing skills or will new, more technical maintenance procedures be required? In one example, a maintenance supervisor announced to her crew, "I understand the new hydraulic excavator has an onboard computer." Considering that the same supervisor is on record as saying, "Over my dead body will I ever use a computer," a maintenance organization like this could present a real "cultural" problem to equipment replacement decision makers. However, although this example is extreme, it is often the reality. If a trend toward using "parts changers" rather than professional diagnosticians has developed in any industry, the need for basic skill training may preclude acquiring new equipment altogether.

Modern equipment will require the application of advanced condition monitoring, the use of the latest technical procedures, and the development and use of quality information. Thus, those involved in the decision to acquire new equipment should look carefully at the current capability of maintenance and its potential to improve.

New tools and procedures—Will a significant amount of new tooling or the major revision of maintenance procedures be necessary too maintain the new equipment? Maintenance personnel may require the use of new tools which then translates into both more training and an additional capital outlay to acquire them. Some of these tools may be more sophisticated diagnostic or condition-monitoring tools. Some may be the upgrading of the information system to be able to download and analyze the data generated by onboard computers. Some maintenance organizations may even have been operating without a modern electronic work order system. Thus, changes and needs are even more complex as new equipment is acquired.

A TEAM REQUIREMENT

The act of replacing equipment cannot be the unilateral effort of accounting personnel with only occasional input from a senior manager. In the new century and as we prepare for its challenges, equipment replacement must be a team effort with field personnel playing a prominent role along with managers and financial personnel. A total picture of requirements, capabilities, limitations, and needs must all be embodied in the decision. Field personnel bring practical considerations into play that better ensure the productive future utilization of the equipment.

ASSEMBLING THE DATA

Tables 16-2 through 16-7 illustrate the basic data collection in the first phase of analysis to acquire new equipment.

TABLE 16-2 Depreciation costs

Data	Year 1	2	3	4	5
Trade-in value, %	75	60	50	40	35
Trade-in value, $	300,000	240,000	200,000	160,000	140,000
Yearly depreciation, $	100,000	60,000	40,000	40,000	20,000
Cumulative depreciation, $	100,000	160,000	200,000	240,000	260,000
Cumulative hours	2,000	4,000	6,000	8,000	10,000
Depreciation cost/hr, $	50.00	40.00	33.33	30.00	26.00

TABLE 16-3 Investment costs

Data	Year 1	2	3	4	5
Investment, $	400,000	300,000	240,000	200,000	160,000
Value at end, $	300,000	240,000	200,000	160,000	140,000
Average investment, $	350,000	270,000	220,000	180,000	150,000
Investment cost (12%), $	420,000	32,400	26,400	21,600	18,000
Cumulative cost, $	42,000	74,400	100,800	122,400	140,400
Cumulative hours	2,000	4,000	6,000	8,000	10,000
Cumulative investment cost/hr, $	21.00	18.60	16.80	15.30	14.00

TABLE 16-4 Maintenance costs

Data	Year 1	2	3	4	5
Availability, %	95	93	90	88	85
Maintenance cost, $	20,000	25,000	40,000	40,000	55,000
Cumulative repair cost, $	20,000	45,000	85,000	125,000	180,000
Cumulative hours	2,000	4,000	6,000	8,000	10,000
Cumulative repair cost/hr, $	10.00	11.20	14.20	15.70	18.00

TABLE 16-5 Downtime costs

Data	Year 1	2	3	4	5
Availability, %	95	93	90	88	85
Hours not available	100	140	200	240	300
Rental cost/hr ($230), $	23,000	32,200	46,000	55,400	69,000
Cumulative downtime cost, $	23,000	55,300	101,200	156,600	225,600
Cumulative hours	2,000	4,000	6,000	8,000	10,000
Cumulative downtime cost/hr, $	11.50	13.80	16.90	19.50	22.60

TABLE 16-6 Obsolescence costs

Data	Year 1	2	3	4	5
Obsolescence factor, %				20	20
Hours required for production				400	400
Cost at $230				92,000	92,000
Cumulative cost, $				92,000	92,000
Cumulative hours	2,000	4,000	6,000	8,000	10,000
Obsolescence cost/hr, $				11.50	18.40

TABLE 16-7 Summary of cumulative costs per hour

Data	Year 1	2	3	4	5
Depreciation cost/hr, $	50.00	40.00	33.33	30.00	26.00
Investment cost/hr, $	21.00	18.60	16.80	15.30	14.00
Maintenance cost/hr, $	10.00	11.20	14.20	15.70	18.00
Downtime cost/hr, $	11.50	13.80	16.90	19.50	22.60
Obsolescence cost/hr, $				11.50	18.40
Totals/hr, $	92.50	83.60	81.20	92.00	99.00

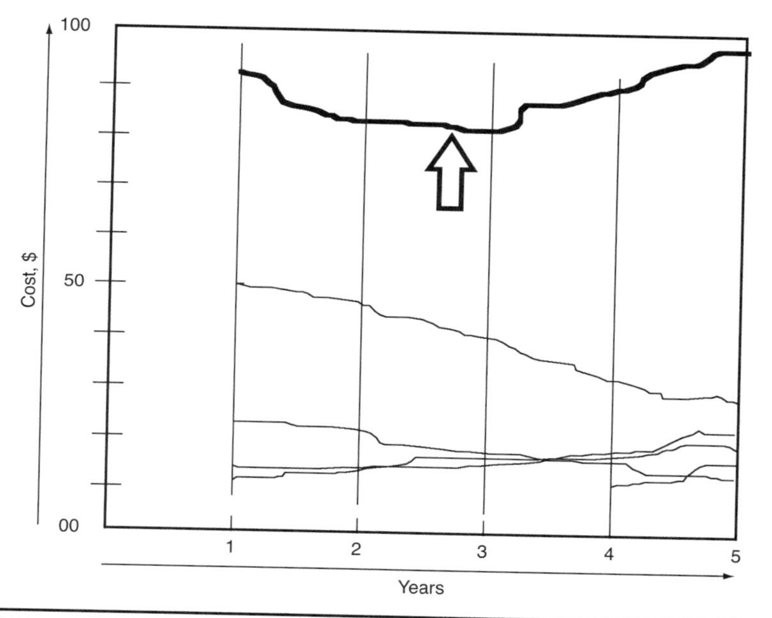

FIGURE 16-4 The cumulative cost of depreciation, investment, maintenance, downtime, and obsolescence dictates the replace of equipment between the second and third years

These data trace the costs per hour of depreciation, investment, maintenance, downtime, and obsolescence for a $400,000 unit of equipment over a 5-year period. The data indicates that the unit should be replaced during the third year when the total of all the factors is at the least cost level but starting to rise. See Figure 16-4.

TABLE 16-8 Equipment replacement considerations

Department	Considerations
Maintenance	Training needs
	New maintenance techniques
	Specialized tools
	Adequate facilities
	More information
	Available technical support
	Ease of maintenance
Operations	Operator training needs
	Productive capacity
	Ease of operation
	No hazards or safety problems
	Cost of operation
	Dependability
	Utilization of equipment
Warehousing	Consignment stock
	Local remanufacture
	Stock replenishment
	Cost
Purchasing	Shipping time
	Component rebuilding support
	Competitive cost
	Warranties
Accounting	Operating cost
	Maintenance cost
	Depreciation
	Cost of obsolescence
	Acquisition cost
	Resale value
	Cash discount
	Tax considerations
	Financing
	Overhaul versus replacement
	Inflation
Safety	Safe operation
	Safety standards met
Environment	Environmental standards met

POLICIES AND PROCEDURES

Certain policies govern acquiring new equipment and making decisions for its utilization, maintenance, and operation. Management should provide for equipment replacement. Then individual departments can develop standard procedures for making recommendations that are important to them. A company team effort is necessary. Although the overall policy can be a single statement, procedures must cover all aspects of economic and performance considerations. Consider the following policy statement:

> Replacement of major production equipment will be preceded by a full justification of the need for the new equipment based on an analysis, the alternatives, and their potential benefit in terms of profitability and performance.

Table 16-8 provides a summary of the principal considerations of each department as their input is sought during debate leading up to the actual replacement decision.

SUMMARY

Maintenance cannot operate effectively without the support of material control departments, including warehousing and purchasing, plus the support of shops that machine, weld, fabricate, and assemble the materials maintenance must install. Equipment management brings all of these departments together by ensuring that departmental interactions are mutually supporting. Maintenance is fully conversant with material control procedures and accountabilities, and in turn, material control departments are well-grounded on the specific ways they must support maintenance. The result is a close working relationship and successful maintenance.

Competition among industries will be more globally oriented in the 21st century. Thus, every industry will be examining how they can become more profitable. Increasing the productivity of equipment will be among the logical steps. In turn, older, less productive equipment will be replaced. Modern production equipment will be expensive as well as more complex. Thus, as new capital outlays are considered, there must be more careful consideration than ever before. The singular economic approach to replacing equipment involving primarily accounting personnel must give way to a company team approach. To ensure that the best equipment for the job at the best price is obtained, economic as well as performance considerations must be weighed. Managers will provide policies allowing each department to recommend specific features of the new equipment. Then, as decisions are made, there will be better assurance that the right choices are made and that the equipment will effectively support the productivity and profitability requirements of the future.

CHAPTER 17

Cost Control

CONTROLLING MINING MAINTENANCE COSTS

Several facts about mining maintenance cost control must be taken into account to achieve successful cost reductions:

- Maintenance is the largest controllable cost in most plants and mines, often representing more than 35% of operating costs.
- Maintenance labor exists to install materials. Thus, the efficiency with which maintenance personnel install materials is a primary factor in their ability to control costs. Therefore, effective utilization of labor is a primary cost-control factor.
- Solid planning is necessary to enable personnel to install materials effectively and reduce or control costs.
- Good planning is only possible if there is a quality preventive maintenance (PM) program. It permits maintenance to uncover deficiencies in time to plan.
- Managers know that the cost of maintenance will not go down unless maintenance is done with fewer people and done less often. Thus, smaller, more efficient maintenance organizations are a requirement for profitability.
- Fewer people often means downsizing.
- Doing maintenance less often means a better maintenance program.
- A better maintenance program requires the diligent, skillful application of the most effective techniques and the use of quality information.

Profitable operations in the 21st century will have successfully downsized, applied the most effective maintenance techniques, and developed and used quality information skillfully. Equipment management provides a guideline for achieving all of these cost-reduction objectives along with delivering reliable equipment.

A FUNDAMENTAL OF COST REDUCTION

Assuming a fixed work force level, reducing the cost of labor to install material is the best way to reduce overall maintenance costs. This requires a strong detection-oriented PM program combined with good planning and scheduling. As PM inspections and testing uncover equipment problems sooner, problems will be simpler and less costly to correct. But these PM services will also find the problems far enough in advance of failure to allow good planning to take place. Thus, every job that is planned and scheduled has the potential for being carried out more efficiently. That is, it will cost less in labor to install each dollar of material. The resulting improvement in productivity will then make labor available to do other work, possibly that work now done by contractors. Herein are the savings. This fundamental relationship is essential to effective maintenance cost reduction.

BUDGETING MAINTENANCE COSTS

Budgeting is the determination of the cost of 1 year's maintenance activity broken down into measurable elements. Yet to be effective, the budget must be used as a management tool. Therefore, the budget becomes

- A plan of action for 1 year's maintenance activities;
- A means of allocating resources to specific maintenance activities; and
- A plan for controlling the rate, timing, and direction in expending allocated funds.

The maintenance budget is a recommendation for the approval of funds to support maintenance activities. It is also a recommended plan of action for the expenditure of funds. A cost-reporting system provides a means of reporting expenditures with provision for the reallocation of funds or the allocation of additional funds for unexpected needs. Typically, maintenance budgeting techniques include

- Zero-base budgeting,
- Performance budgeting, and
- Factored budgeting.

Budgeting is an effective maintenance management tool in which maintenance makes a commitment to accomplish their program with specific funding levels. In turn, budgeting procedures control the rate, timing, and utilization of funds while providing management oversight to monitor and manage the plan as well as its accomplishments and money expended.

Zero-Base Budgeting

The zero-base budgeting technique begins by establishing activities that will be subjected to budgeting controls. Then objectives are set for these activities, with standards and performance methods established to meet the objectives. Next, priorities are set for funding the selected activities, and the activities compete with one another for funding. Thus, only the most important are funded and performance is measured on each. A continuing review procedure identifies programs that do not meet objectives so that funding can be cut back or eliminated. Thus the name "zero-base," with the result that attention is drawn more quickly to essential activities.

Performance Budgeting

The performance budgeting technique involves the allocation and control of resources through work activities. The activities are defined so that maintenance personnel can plan, schedule, control, and report on the work they do. Work activities must lend themselves to work–cost evaluation. Consequently, each significant activity should also relate to standardized resource requirements of labor, equipment use, material consumption, time required, and cost. Maintenance must be able to break down each significant activity into resource requirements and relate them to production rates and costs.

Consider the following breakdown for janitorial work such as power sweeping of plant aisles and working areas:

Unit of measure:	1,000 ft^2
One man per day:	60,000 ft^2
Man-hours per 1,000 ft^2:	0.133
At \$5.20/hr, the cost of operating the sweeper:	\$7.00/hr
Sweeping cost per day (60,000 ft^2):	\$97.60
The cost per 1,000 ft^2:	\$1.63

Then work level is specified (how much sweeping will be done?), and the work load is computed: 60,000 ft^2 swept weekly × 50 weeks = 3,000,000 ft^2/yr. The work load is then converted to resource elements:

Man-hours:	399
Labor cost:	\$2,075.00
Equipment cost:	\$2,793.00
Material cost:	\$0
Total cost:	\$4,868.00

The weekly work performed is then evaluated against performance targets:

Activity:	mechanical sweeping
Work measurement:	1,000 ft^2
Rate:	33 man-hours/ft^2
Cost:	\$1.63 per 1,000 ft^2

A performance evaluation is possible on a periodic basis. Consider a typical monthly report for July:

Annual work quantity (ft^2):	3,000,000
July	300,000
To date	1,800,000
Remaining	1,200,000
Annual labor requirement (man-hours)	399
July	40
To date	239
Remaining	160
Performance (man-hours/1,000 ft^2)	
Standard	0.133
July	0.133
To date	0.132
Total cost (budget)	$4,868.00
Period	$489.00
To date	$2,934.00
Remaining	$1,934.00

Performance reports similar to this show the performance status (0.132 to date versus a standard of 0.133) as well as the costs accrued against the budget.

Factored Budgeting

The factored budgeting technique relates the use of maintenance man-hours to production targets and then converts the relationship to labor cost and material cost to carry out the maintenance program. When supervisors are given a ready-made budget and told to "live with it," they have little incentive to achieve budget objectives. Therefore, maintenance cost control should initiate the budgeting process at the supervisor level to ensure commitment. Because budget goals must be met, one way a maintenance manager can involve supervisors and team leaders while encouraging them to meet budget goals is through the use of factored budgeting. This technique offers a means of preparing a realistic program to control maintenance costs by relating equipment repair history to supervisor judgments. With factored budgeting, the maintenance manager can tie together maintenance data and production targets.

Factored budgeting is a means of budget preparation that provides a plan of action. Every area of maintenance cost is logically considered. Primarily, it is a recounting of the number of man-hours that have been needed to maintain specific key machines. In this way, desired production levels can be sustained. Labor estimates are stated in man-hours for a true picture of the work load. The

TABLE 17-1 Maintenance budgeting responsibilities

Requirement	Action		
	Planner	Supervisor	Manager
Review man-hours used to maintain equipment	X		
Note changes in equipment volume		X	
Determine extraordinary maintenance requirements		X	
Project man-hours required	X	X	
Determine material requirements	X	X	
Convert man-hours to labor dollars	X		
Distribute costs	X		
Prepare budget plans	X	X	
Review and approve budget			X

historic relationship of man-hours to production output establishes the ratio of results (e.g., man-hours/ton). Applying these ratios to desired future production levels will ensure an accurate projection of future man-hour requirements. And, if average wage rates are used with projected man-hours, labor costs can be easily determined as well. In most plants, the maintenance supervisor develops the man-hour work load and material requirements. He or she must also determine when extraordinary expenditures are necessary. Some of the activities devoted to developing maintenance plans are shown in Table 17-1.

As the preliminary equipment repair projections are made, a ratio between the number of man-hours required for maintenance and the amount of product generated by the equipment to be worked on should be considered (tons, boxes, etc.). In addition, the following considerations are pertinent:

1. Which units of equipment will require extraordinary attention (major component replacements, rebuilds, or overhauls)?
2. During which month/quarter should this extraordinary action be planned?
3. What resources are necessary for the maintenance of items that are not identified individually (other)?
4. How much overtime should be budgeted, and where can it be used most effectively?
5. What will be the effect of vacations, sick leave, and absences on the annual maintenance plan?
6. Where will contract work be required, and when should it be scheduled?
7. How much more will materials cost next year, and how will maintenance costs be affected by inflation?

TABLE 17-2 Basic maintenance history data

Asset No.	Description	Man-hours	Labor Cost	Material Cost	Total Cost
128171	Kiln 1	2,131	$9,806	$16,699	$26,505
128172	Kiln 2	2,146	$9,874	$20,983	$30,857
128173	Kiln 3	3,757	$17,280	$30,720	$48,000
128174	Kiln 4	2,157	$9,920	$22,080	$32,000
151990	Other	7,332	$33,727	$59,959	$93,686

TABLE 17-3 Adjusted projections

Asset No.	Description	Man-hours	Labor Cost	Material Cost	Total Cost
128171	Kiln 1	2,258	$10,390	$47,699	$58,089
128172	Kiln 2	2,274	$10,460	$21,612	$32,072
128173	Kiln 3	2,100	$9,660	$16,665	$26,325
128174	Kiln 4	2,286	$10,515	$17,500	$28,015
151990	Other	6,130	$28,198	$31,950	$60,148

When these questions are answered, the supervisor can begin to develop specific aspects of the budget necessary to meet budgetary goals. A typical format for a preliminary budget study might contain details similar to those listed in Table 17-2.

After reviewing the data in Table 17-2, extraordinary maintenance needs are considered:

- Kiln 1: A new ring required. The cost will be approximately $31,000, included under materials.
- Kiln 3: Costs reflect a major rebuild that took place last summer. Revise costs downward to show fewer costs.
- Area responsibilities were changed. This results in a reduction in "other," as other supervisors will do it.
- A 3% increase in material costs will push all material costs upwards.
- The department target for overtime will be 6%, down from 8% last year.

These adjustments are then applied, resulting in Table 17-3.

Historical trends of relatively stable maintenance activities such as lubrication, PM inspections, running repairs, and so forth, should be reviewed to determine if there is a cost increase as a result of adding new equipment or using new condition-monitoring equipment versus PM services done by personnel.

Then, as soon as the supervisor has completed preparation of the basic man-hour and cost figure data, these should be checked with the manager. If the manager agrees on the accuracy of the data, a cost distribution plan is developed next. A uniform monthly distribution of costs is not always possible

TABLE 17-4 Distribution of costs by month

January	0.8299	July	0.8064
February	0.8496	August	0.8121
March	0.8496	September	0.8467
April	0.8467	October	0.8323
May	0.8380	November	0.8496
June	0.8121	December	0.8409

because of vacations, sicknesses, holidays, and so on. Unusual heavy operating expenses (e.g., major repairs) should be budgeted several months apart so that their cost impact is reduced. A typical distribution of costs is indicated in Table 17-4.

A key element in the factored budgeting technique is the ability of the supervisor to assess the current status of the equipment and to evaluate special needs for the future, especially for the coming year. The supervisor is guided by such historical data as man-hour use, man-hours/unit of product ratios, and expenditures in previous years. These data also highlight cyclic needs such as overhauls or rebuilds, for instance. They also indicate whether these major activities will be necessary during the coming year. The supervisor's participation in factored budgeting preparation is essential, as he is closest to the overall maintenance picture and is the one most familiar with the equipment. Furthermore, when he participates in budget development, he is also committed to carrying it out. Most current budget techniques are simple to use with computer applications that are readily available. Historical data and projections are entered as prompted by programs. Once developed, regular cost accounting takes over.

STRATEGY OF COST REDUCTION

Many plants believe that all maintenance and repair work is necessary. They further believe that all work must be done as soon as it becomes known. The typical maintenance supervisor is primarily service oriented rather than evaluation oriented. Surprisingly few maintenance supervisors question whether work should be done. They rarely question the personnel bringing work to them as to how essential it is. In a sampling taken of ongoing work in several machine shops and fabrication shops, it was found that

- Thirty-eight percent of the work was brought directly into the shop by operating personnel and started without approval by production supervision.
- Sixty-seven percent of all work had no priority rating and was being handled on a first in–first out basis.

In three other instances, the following situations occurred:

> An approved weekly schedule was evaluated to determine how closely it was being followed. The schedule compliance revealed that only 31% of the items on the schedule were attempted and only 17% completed. A closer examination of the underlying reasons showed that the foremen themselves accepted sufficient other unapproved work to use up more than 60% of their available labor (an example of "good will" work). Thus, they themselves made decisions that precluded their compliance with the schedule.
>
> Another plant consistently worked on more than 62% of major work orders that were still being planned and had not yet appeared on an approved schedule. Foremen were provided blueprints in advance only to study them. Despite the fact that they were not yet authorized to do this work, they did it instead of the work actually scheduled. Further questioning about this incident revealed that these foremen released the blueprints prematurely with no instructions to their workers. Thus, the impetus for starting the work came from the craftsman level. The foreman had not even bothered to try to control the work.
>
> In one surprising case in a large plant, three general foremen participated in a working group to develop a priority system. They approved its wording and signed a document recommending its adoption by plant management. It was approved and published. Ninety days later, 12 of the 17 foremen working for these general foremen had not even heard of the priority system.

It seems obvious that closer work control procedures must not only be developed, but steps must be taken to ensure they are followed. The following general situation seems to permeate the lower end of the maintenance organization as well:

> In several plants, service tickets were prepared by supervisors and placed in slots that announced these tickets represented work to be done. Because no guidelines or priorities were ever provided, workers helped themselves to the most desirable jobs or the jobs that ensured them overtime work. In several instances, some workers placed jobs in another worker's slot and helped themselves to jobs assigned other workers.

Unfortunately, too many instances of this type take place too frequently in many maintenance departments. Such lack of work control can easily undermine serious efforts in the reduction of maintenance cost.

In plants that have successfully reduced maintenance costs, they have achieved lasting cost reductions by insisting that maintenance establish controls and use them. Successful cost reduction is predicated on work control. Consider these useful "ground rules":

- Maintenance work should be approved before it is performed.
- Activities such as PM services should be approved by operations in "concept" and then performed automatically.
- Major jobs should be cost-evaluated and approved by a supervisory level consistent with the value involved.
- Policies should preclude performance of good will work.
- There should be a criteria identifying work requiring planning.
- Planned work should be prioritized.
- Planned, prioritized work should be scheduled.
- Scheduled major work should be subjected to labor allocation. It should also be rejected if unnecessary and deferred if not timely.
- Use of labor should be scrutinized, performance measured, and questionable use justified by supervision.

There are practical ways to follow these ground rules:

- The PM program should be approved in total scope before it is started.
- A means for monitoring PM compliance should be established.
- The work order procedure should contain a mechanism for estimating the cost of the work.
- Cost levels should then require approval by specific supervisory levels.
- No work should be approved by staff personnel such as planners.
- Only line supervisors should approve work.
- A substantial penalty should be imposed for doing good will work.
- Every major job should be screened to determine if it needs to be planned.
- Work should never be released for scheduling until all planning is completed.
- Planned work should be prioritized to guide the allocation of labor.
- Scheduling should always include production in decisions on labor allocation, deferring work, or rejecting work if not considered necessary.
- Labor use should be carefully measured and justified.

Plants that have followed these general guidelines conscientiously have identified significant improvements:

- A foundry experienced an immediate 28% reduction in its backlog.
- Two reduction plants were able to form a craft pool with surplus field labor. Then they allocated labor from the manager level. They soon achieved a 16% reduction in the work load and were able to reduce the work force through attrition without having to replace personnel.

- A paper mill determined that 12% of its planned work was given too high a priority and 6% was not necessary at all.

All of these examples have demonstrated that work control is essential to cost reduction.

AVOIDING ARBITRARY COST REDUCTION STEPS

The dramatically increasing costs of maintenance are causing great concern among managers. In the past, managers have been forced to make arbitrary reductions in the maintenance work force to slash costs. As production rates resurged, these managers found that their arbitrary actions had undermined maintenance capabilities. In retrospect, the more appropriate action would have been to determine more precisely where maintenance costs could be reduced and aim the cost reduction effort there. The result would have been a cost reduction in the right areas with the emergence of a better maintenance effort rather than a disabled maintenance department.

REVISITING THE FUNDAMENTALS

Senior managers are getting more concerned about the cost of maintenance as the competitive posture of competing industrial plants dictate an even greater effort required in the 21st century. Most are keenly aware that maintenance costs are difficult to reduce unless the maintenance is carried out with fewer people and less often.

Regrettably, few senior managers have had little exposure to managing a maintenance department. Therefore, they are less certain of how to cause maintenance to be done less often. As a result, they favor downsizing maintenance as the surest way they know to reduce costs.

A first management-level action to reduce cost should target better maintenance done less often and then improve productivity to get greater value from labor. These steps, in this order, will more surely produce a surplus of labor over work load, permitting a gradual reduction in the work force through attrition.

Suppose that as a senior manager, one could inspire maintenance to implement reliability centered maintenance (RCM). If it did, it would apply condition-monitoring techniques to replace the limited value inspection and testing formerly done by human beings. As a result, problems that human beings were incapable of detecting would be identified accurately using condition monitoring instead. In turn, components that used to be changed at the end of a predetermined time period could be left in service longer and monitored to ensure they were functioning effectively. Thus, the introduction of RCM and the use of condition monitoring would net an increased life-span of most components.

It is with techniques like this that maintenance can be done less often. Thus, a manager wishing to reduce costs should consider the principles of equipment management as they apply to equipment: new technology and better information. Next, the people part of cost reduction must be put into focus. Productivity must be improved to set the stage for using fewer personnel to conduct maintenance. Here are a few guidelines to make this happen:

1. Insist that productivity be measured.
2. Verify that the maintenance program has been well-defined and that all plant personnel understand it.
3. Ensure that the work load is carefully determined (the right number of people of the proper craft).
4. Leaving supervisors with only enough labor for day-to-day work (preventive maintenance, unscheduled repairs, emergency work, and routine tasks), establish management-level controls to allocate the remaining labor (with sufficient supervision) by priority to major, periodic tasks such as plant shutdowns, major component replacements, or overhauls.
5. Establish a moratorium on nonmaintenance engineering project work (if this work is done by maintenance and there is no construction crew).
6. Have senior managers attend all maintenance scheduling meetings and insist on a report of schedule compliance while carefully watching the utilization of labor (more preventive maintenance will yield more labor for planning while reducing the labor required for fewer emergency repairs).

If these guidelines are followed, the promised excess of labor over work required will be at hand. Now, at a leisurely pace, attrition can be applied to gradually reduce the work force levels—and the cost of maintenance. The legacy of the experience will produce a work force that now more than meets the required work load with fewer people while doing the work less often.

COST-REDUCTION GUIDELINES

The cost of maintenance is made up of the direct cost of performing work and the indirect cost of downtime, resulting in lost production when inoperative equipment is not producing. Direct costs approximate 35% of total operating costs. Indirect costs add 300% more when missed production results in marketplace losses. Therefore, mining managers must take steps to reduce maintenance costs. Although the equipment management strategy and supplements like RCM provide a direction for cost-reduction efforts, actual achievement centers on two harsh realities. The cost of maintenance will not go down unless it can be done with fewer people and done less often. Only the application of practical, realistic steps will ensure that these realities are achieved. This section outlines these steps.

If maintenance is to be done with fewer people, the organization must either downsize or improve productivity. Downsizing will result in fewer people. But if done carelessly, it can undermine the capability of the organization to accomplish basic maintenance. It is better to improve productivity as a means of using fewer people. Then when productivity has been improved, reductions can be made more accurately and deliberately, knowing that performance has not been undermined. If maintenance is to be done less often, the organization must apply the principles of equipment management supported by the best technology and effective use of the information.

The first step in improving productivity is to verify and document the maintenance program and to ensure that it parallels the principles of equipment management. The maintenance program must also coincide with the information system that supports it.

The next step is to verify the work load as the essential work performed by maintenance. In doing so, this step also identifies the right number of personnel by craft required to carry it out, working at a reasonable level of productivity.

Next, the roles of key personnel must be defined and verified and carefully matched with the program elements.

Then maintenance terminology must be defined to eliminate any confusion within maintenance or between departments as they support maintenance.

The education of key maintenance, operations, staff, and management personnel is a big part of the success of equipment management, and the effort to reduce costs and ensure profitability. This education centers around the maintenance program—its clarity, and how well it is understood and applied. When a maintenance program is well-defined, individuals will know exactly what should happen and how, leading to better overall performance. Effective program definition also ensures that the best organization is established and the right information is identified and effectively applied. Program definition is an opportunity to think through the details. But the real value results when maintenance, operations, and staff departments debate and discuss the details to arrive at the best program. Education is also an effective way to prepare personnel to support the program and accomplish desired improvements.

With regard to doing maintenance less often, a more vigilant preventive maintenance program with the diligent application of modern condition-monitoring technologies will ensure that most equipment problems are uncovered far in advance of equipment failure. In turn, fewer major repairs will be required. But those that are required will be well planned and accomplished with less cost, in less time, and with a higher level of worker productivity.

Finally, identify areas needing improvement. The same evaluation procedure can also be used to assess progress in cost reduction, especially in the detail of whether maintenance is actually being carried out more productively (with fewer people) and more effectively (less often).

SUMMARY

Genuine, lasting maintenance cost reduction must be a plantwide effort. By itself, a maintenance organization with a solid program can attempt cost reduction. But without the full support and cooperation of all other departments, maintenance cost-reduction achievements will be limited. Equipment management ensures that operations will use maintenance services effectively and material control departments will support maintenance with quality service. Mutually-supporting efforts such as these, carefully implemented with equipment management, will yield the desired cost reduction that is so necessary to achieving plant profitability. One of the major themes of mining cost reduction is the acknowledgment that maintenance costs are unlikely to go down unless maintenance can be done less often and with fewer people.

CHAPTER 18

Assessing Maintenance Performance

EVALUATION: THE FIRST STEP OF IMPROVEMENT

The 21st century has fostered intense competition among industrial plants. Only those plants that can achieve and sustain profitability will survive. In the process of survival, these plants will also have developed an effective maintenance effort, one that controls and reduces costs while ensuring maximum equipment reliability. These plants are also the ones that never guess at their performance and profitability status. They always know where they stand because regular, continuous evaluations are an integral part of their operating strategy.

PURPOSE OF EVALUATIONS

An evaluation establishes the current performance level by identifying those activities needing improvement as well as those being performed well. The evaluation not only confirms performance but is the starting point for any improvement effort. Yet the most important by-products of a well-conceived and effectively conducted evaluation are the education of personnel and their commitment to provide genuine support for improving performance. An evaluation that successfully points the way to improvement makes it more attainable and is a matter of pride to the organization. Thus, the very people who can create the improvements are ready to help in a cause they already believe in.

OVERCOMING RESISTANCE TO EVALUATIONS

Resistance to evaluations is often based on misunderstanding and suspicion. But maintenance organizations usually resist evaluations because there are too many factors over which they have no control that influence their performance. Thus, a poor maintenance organization may hope to delay an evaluation by issuing misleading "data" or warning of unpleasantness from workers. Under such circumstances, a plant manager might not go ahead with an evaluation. Thus, problems that demand to be corrected are never identified. Typically,

- The warehouse may be poorly run because they are uncertain of the exact support required by maintenance.
- Operations personnel may not make equipment available for essential maintenance because they are pushed into meeting unrealistic production targets.

In both of these instances, maintenance is the victim. There is little it can do, by itself, to change the circumstances. Management must eliminate resistance by removing any threat to maintenance implied by an evaluation. Therefore, the evaluation must distinguish between those factors over which maintenance has responsibility (e.g., quality work) and those over which it has no control (an unresponsive warehouse).

Then if poor maintenance performance is due to inadequate warehouse support, the responsibility of the warehouse to support maintenance must be clearly defined and its performance assessed. Only then can it be corrected. In turn, maintenance can only be evaluated fairly if other departments whose actions impact its performance are assessed as well. After it becomes clear that the factors over which maintenance has little control will also be evaluated, most resistance to evaluations will be reduced.

MAINTENANCE LEADERSHIP

The limited management background of some key maintenance leaders can also influence their willingness to be evaluated. Typically, maintenance supervisors and even superintendents are promoted from within maintenance. Often they carry their management styles upward, largely unchanged. If they are ex-craftsmen, their "school of hard knocks" backgrounds have taught them how to make things run again. But they may not know how to manage resources to maintain equipment plantwide. University degrees, though not a requirement to manage maintenance, are rare among maintenance leaders when compared with operations, for example. Few supervisors have had significant management training. They often assume their supervisory responsibilities with little preparation for the challenges they will face. Regrettably, too many managers in the past have placed little value on educating maintenance supervisors. Rather, they considered it sufficient that they simply "got the equipment running again." Less attention was paid to an education that would equip maintenance leaders to establish a viable program to preclude poor maintenance performance. As a direct result of these omissions, some supervisors may be unwilling and even uncomfortable in their work control roles. For these reasons, plant managers, in preparing to launch a strategy that will include regular, periodic evaluations, must target maintenance supervision for education on the purpose and objectives of the intended evaluations.

SUCCESSFUL EVALUATION STEPS

An effectively conducted evaluation can identify needed improvements and assign priorities to corrective actions. But the evaluation can also provide the opportunity to initiate corrective actions. The evaluation is the first of step toward improvement. It provides the "as is" status of current performance. However, the more important aspect of the evaluation is to successfully convert the results into a plan of action and exert the leadership to implement needed improvements. Ten steps are involved.

Step 1. Develop a policy for evaluations. There must be a management policy requiring that maintenance be evaluated on a regular basis. Such a policy will preclude the resistance that many poorly run maintenance departments exhibit to discourage evaluations. This policy will help to redirect the energy of resistance into efforts to prepare for evaluations. A typical policy by a plant manager might be as follows:

> Maintenance makes a significant contribution to the overall profitability of the plant. To ensure that maintenance services are effective, we will evaluate maintenance and the actions of other departments that support or cooperate with maintenance. Evaluations will be carried out on a regular, continuous basis. The evaluations will examine how effectively we perform our work, as well as how the plant and its operating and staff departments support and utilize maintenance services. Evaluation results will be publicized so that we may know where we stand, what we have done well, and where we need to improve. Evaluation results will be publicized immediately, and they will serve as the basis for an improvement plan to be implemented through the cooperative efforts of every plant department. We look forward to your continued cooperation.

Step 2. Provide advance notification of the evaluation. Advise personnel about the evaluation and make a preliminary statement about its content, purpose, and use of the results. Eliminate surprises with the announcement but emphasize the policy of regular, continuing evaluations. The announcement should

- Advise of evaluation dates,
- State the objectives,
- Summarize previous evaluation results,
- Restate improvement objectives,
- Thank everyone in advance,
- Advise that their help is critical to improvement, and
- Offer to discuss announcement details.

Step 3. Educate personnel on the purpose of the evaluation. Explain that the evaluation is a checklist describing what should be done. Results describe how well they did and provide a basis for improvement. They identify what is done well, not what is done poorly. The evaluation is an opportunity to help maintenance account for its contribution to plant objectives. Emphasize the positive aspects of the evaluation through education.

Change unfavorable misconceptions about evaluations by telling personnel that the evaluation is a means of finding out how they can do better. When preparing personnel for the evaluation, acknowledge that there may be a genuine fear of audits and evaluations. This fear may be based on previous bad experiences, or it may have no basis at all. Many maintenance workers simply don't want anyone "looking over their shoulders." Other fears are fueled by managers who themselves may create uncertainty with remarks that mislead. One maintenance manager, for example, proclaimed that "auditors are those who come in after the battle is lost and bayonet the survivors!" This type of cavalier attitude contributes little to the mental preparation that must be made if a plant is struggling to improve. Some worker groups have been known to threaten slowdowns or strikes when an evaluation is announced. They fear that their "sacred promise" of preserving jobs will be threatened by possible work force reductions as the result of an evaluation. The fact that the evaluation can also lead to improvements that will make their work easier may not occur to them. It follows that maintenance supervisors, caught up in the threat of their jobs being made harder by the union position, may resist as well. Thus, education becomes doubly important. Education also avoids surprising personnel, as this adds to resistance and creates distrust. People don't like surprises. Therefore, educate personnel about what is coming, which helps avoid resistance.

Resistance to being evaluated also occurs when there are several operations that will be evaluated. The concern is that performance between several operations will be unfairly compared when, in fact, they are not similar and would be better compared on another basis. Regardless, comparisons will be made. The sponsor of dual evaluations, usually a general manager, must convey a supportive attitude. The sponsor should provide encouragement to conduct the evaluation and follow up to see that something constructive is done with the results. If help beyond the resources of a single plant is necessary for improvement (e.g., computer programming), the sponsor can help strengthen good relations by providing it.

Education continues through the evaluation process and into the results. In individual plants, for example, local managers will want to know how well their policies are understood and how effectively the procedures based on those policies are being carried out. Although they are concerned with the quality of the maintenance program, they will be equally interested in learning how well, for example, production supports and cooperates with the program. Therefore, in a multiplant environment, managers will be less concerned with what others may think. Instead, they should assume that every other plant is interested and concerned but also ready to help them rather than compete with them.

Step 4. Schedule the maintenance evaluation. Schedule the evaluation carefully, particularly if it will involve a physical audit lasting several weeks. In selecting the evaluation dates, be aware of potential conflicts that might distort evaluation results. For example,

- If a shutdown has just been completed, there may be distracting startup problems.
- Similarly, if a shutdown is coming up, preparation for it may compete for the attention of personnel in getting ready for the evaluation.

Peak vacation periods may also find key personnel away from the plant. Be aware of the absences of key people and weigh their nonparticipation in deciding when the evaluation should be conducted. Similarly, recent personnel changes could limit knowledge of evaluation points, and personnel cutbacks or staff reductions might affect attitudes. In general, the evaluation should be carried out in a stabilized situation with as few distracting conditions as possible. With suitable advance notice, the organization can prepare for the evaluation and look forward to learning how they are doing.

Timing of evaluations is also important. When they are conducted on a regular, continuing basis, people will be constantly preparing; thus, performing better. In addition, they will look forward to the evaluation as an opportunity to demonstrate their progress. Generally, if maintenance personnel feel that the evaluation is constructive, they will prepare for it without hesitation. In subsequent evaluations, if they have accepted the evaluations and are convinced of their value, they will make a conscious effort to improve on previous results.

Step 5. Publicize the content of the maintenance evaluation. Make sure the content of the evaluation is explained in advance. A maintenance evaluation covers such a variety of activities that it is unlikely for any advance notice to influence a dramatic shift in performance. By announcing what is to be evaluated, organizing for the evaluation is helped. Reports will be ready, personnel scheduled for interviews will be prepared, and the evaluation can be carried out more effectively. Agreement on evaluation content also clarifies expectations of both the maintenance department and the evaluators. For example, if productivity is to be measured, hourly personnel should be advised so they do not misunderstand the intent. Generally, the scope of most evaluations centers on three broad areas:

1. The organization,
2. The program, and
3. The environment in which maintenance operates.

To evaluate the organization, examine these elements:

- The efficiency in controlling personnel and carrying out work
- The effectiveness of supervision
- Work load measurement

- Labor utilization
- Productivity
- Motivation
- The effectiveness of craft training
- The quality of supervisor training

In evaluating the maintenance program, assess the following areas:

- Program definition
- Use of proper terminology
- The effectiveness of preventive maintenance
- Control provided by the work order system
- The use of information
- Planning quality and the use of standards
- If scheduling yields timely work completion
- Whether work control ensures productivity
- How well maintenance engineering ensures equipment reliability
- Effective use of technology
- The quality of work

In evaluating the environment in which maintenance operates, examine these issues:

- Whether management has assigned clear, realistic objectives
- If guidelines ensure common understanding and compliance
- Management support for maintenance
- Operations understanding and cooperation
- Staff department support and service quality
- Adequacy of material control
- Safety compliance
- Housekeeping quality

The working environment reflects the attitude of the plant toward maintenance. A poor attitude is the most common reason for resistance to evaluations. Even a first-rate maintenance organization with an effective program can fail if the environment does not provide receptivity and support. Therefore, examining the environment must include answers to some difficult questions regarding

- Management sponsorship,
- Operations cooperation, and
- Level of service of staff departments, such as warehousing.

Not surprisingly, when maintenance personnel realize that the evaluation includes key factors over which they have little control but which influence their performance, their resistance diminishes. But be aware that an evaluation of the plant environment could reveal an unwilling operations department or an uncooperative purchasing agent. Therefore, it will be necessary to ask the support of the plant manager in conditioning these other departments for their participation as well as helping to improve their attitudes. Examination of the environment can include steps that the plant manager must take to ensure the success of the program. Typically, these steps include a clear mission and realistic guidelines for other departments to follow in how they must support maintenance.

Step 6. Use the most appropriate evaluation technique. Evaluation techniques should be considered based on the plant situation. Some operations may require an evaluation in which every detail must be scrutinized. Other operations, having established the essential pattern of evaluations, may simply check progress by measuring only a few critical areas. There are three techniques that can be used to evaluate maintenance performance:

1. Physical audit,
2. Questionnaire, and
3. Physical audit with questionnaire.

A physical audit is usually conducted by a team. Generally, the team is made up of consultants or company personnel (or both). They examine the maintenance organization and its program, as well as activities that affect maintenance (such as the quality of material support). The physical audit examines the total maintenance activity firsthand by observing work, examining key activities (such as preventive maintenance, planning, or scheduling), reviewing costs, and even measuring productivity.

The audit relies on interviews, direct observation of activities, and examination of procedures, records, and costs. When done properly, the physical audit produces effective, objective, and reliable information on the status of maintenance. The physical audit by itself should be used if personnel could not be frank, objective, and constructive in completing a questionnaire. But the physical audit is also useful when plant personnel want an opportunity to discuss issues and offer suggestions. However, a physical audit lasting several weeks can be disruptive because of the time required to help explain procedures or participate in interviews. Therefore, it must be well-organized in advance.

A questionnaire gives a cross section of randomly selected personnel an opportunity to compare maintenance performance against specific standards. This cross section might include personnel from management, staff departments (such as purchasing), operations, and maintenance. Although a questionnaire is subjective, the results are, nevertheless, an expression of the views of plant personnel. Therefore, participants will have committed themselves to identifying improvements they see as necessary. To most of them, this constitutes potential support for the improvement effort that must follow. When administering a questionnaire, be careful to ensure that personnel are qualified to respond. For example, participants outside of maintenance must respond only to those standards on which they have personal knowledge. The evaluation must be administered so that questions on evaluation points can be answered completely. When carried out properly, the questionnaire can produce reliable results quickly while minimizing disruption to the operation.

The questionnaire has the advantage of being administered often so that progress against the initial results (benchmark) can be measured. For example, one plant was able to establish areas in which improvement was still needed while setting aside areas in which good progress had been demonstrated by previous evaluations. The questionnaire is the best choice when a quick, nondisruptive evaluation can serve as a reasonable guide in developing an improvement plan. It must be carefully crafted so that it embraces all the elements of the maintenance program, like preventive maintenance, as well as those activities that affect the program, such as purchasing. The appendix, Maintenance Performance Evaluation, illustrates the type of standards against which maintenance should be evaluated. A cross section of about 15% of the plant population, including operations and staff personnel (e.g., accounting), is adequate to produce good results. Participants should have personal knowledge of maintenance performance for the standards against which they are comparing maintenance.

There should be selectivity in who responds to what. For example, the accounting manager could evaluate the quality of labor data reported but could not evaluate work quality. Questionnaires are rarely of value if they are not administered in a controlled environment. Typically, if distributed for completion at the respondents' leisure, expect poor results because there are too many opportunities for misunderstanding when questions about the questionnaire cannot be answered quickly or interpreted by the administrator. Questionnaires administered in a controlled environment, where participants can be oriented and their questions answered, are always best.

The combination of a physical audit and a questionnaire provides the most complete coverage, as the techniques work together. The questionnaire, for example, provides confirmation of physical audit findings. This combined technique is preferred by consultants and corporate teams because it combines the objectivity of the outsider in the physical audit, while the questionnaire helps to educate personnel and gain their potential early commitment to improvement. As the insiders, their help is needed. Without it, little will happen. This combined technique is the best way of preparing for the improvement effort that must follow.

Step 7. Announce results. By publicizing the evaluation results, there will be clear evidence that both the good and the bad are acknowledged. More importantly, there is a commitment to do something. By sharing the results of the evaluation with plant personnel, there is confirmation that their help is expected in attaining improvements. Conversely, keeping the results a secret will decrease credibility and make improvement actions more difficult. Evaluation results should be discussed openly and constructively so that the personnel who must later support improvements are being brought along.

A maintenance superintendent should never rationalize that he was solely responsible when the evaluation reveals a poor performance. He probably got a lot of help from many people in the resulting poor performance. For example, plant management may not have provided the necessary guidelines to ensure that the maintenance program could be carried out in a positive environment. Similarly, operations may have required so many equipment modifications that maintenance was left with inadequate staff to maintain existing equipment. Likewise, supervisors should not be blamed for disappointing results. Little is gained by trying to fix blame.

Maintenance reaches into so many areas that few people would be without some degree of responsibility for the decline of maintenance performance. Simply determine the current performance level and move forward from there. Maintenance can only offer their services, and hope for cooperation and support. Always work toward creating an environment in which support and cooperation can be ensured. By including operations and staff department activities in the evaluation, aspects over which maintenance has little control are assessed and corrections that help maintenance are identified for improvement. Plants performing poorly often do not show results to the personnel who participated. Better plants not only share results but seek help in interpreting the results and soliciting recommendations. For example, one successful maintenance manager observed that "whatever the current performance is, it didn't get that way overnight." He approached his task of improving maintenance performance by saying, "since this is what we think of our maintenance program, then let us now consider what we must do about it." His plant demonstrated that it was already on its way to improving because the support and enthusiasm for doing better was successfully harnessed even before the evaluation was completed. There was sincere involvement.

Most plant management personnel are uncertain of the roles they must play in creating an environment for a successful maintenance program. They acknowledge that stating a clear mission, providing guidelines, and requiring a definition of the maintenance program are useful approaches to the creation of a positive environment for maintenance. Beyond that, uncertainty exists. Thus, a careful review of evaluation results with management will help considerably to support the improvement effort. Production personnel generally do not realize the comprehensive nature of the maintenance program and how much its success depends on their support and cooperation. Therefore, the inclusion of operations personnel in the evaluation and a comprehensive discussion of their potential contribution to improvement is essential.

Staff personnel in purchasing, warehousing, accounting, and information departments often express a need to better understand the maintenance program so they can support it more effectively. Therefore, a discussion with them on the nature of their support will often ensure better service. Many maintenance managers themselves acknowledge that their own maintenance programs are not adequately defined, much less properly explained to either maintenance personnel or the rest of the plant. Some admit a need to align basic program elements such as preventive maintenance, planning, scheduling, and maintenance engineering. Evaluations often reveal facts about the maintenance program that were concealed by defensive information reaching managers or the misleading attitudes taken by maintenance personnel.

Step 8. Take immediate action on the results. The most convincing way to demonstrate that the evaluation was a constructive step is to organize an improvement effort immediately. Obtain commitment for the constructive use of the results by converting them into an improvement plan and immediately organizing the improvement effort. This is the main objective of the evaluation. If the evaluation is one of a series, results should be compared with the previous evaluation. This demonstrates progress as well as the identification of areas that need more work. Separate the good from the bad. Offer congratulations on the good performances, and organize the activities requiring improvement into priorities. Actively solicit help from anyone capable of providing it. Most will participate willingly. Thereafter, develop a plan for further improvement and implement corrective actions. If there are corrective actions beyond the capability of maintenance, don't hesitate to seek help. Plant managers are usually pleased to be asked to help. It is also gratifying to learn that corporate managers, particularly those responsible for multiple plant operations, are eager to help as well. Set up an advisory group and get under way. Let them first determine why certain ratings were poor. Then ask for recommendations for improvement. Change the members of the advisory group frequently to encourage different views. As recommendations are made, try them in test areas before attempting plantwide implementation.

Step 9. Announce specific gains resulting from the maintenance evaluation. As soon as any gains that can be attributed to the evaluation can be identified, announce them and give credit to the appropriate personnel. People like to know how they did. Tell them. In the process, candor will invariably encourage a greater effort in future evaluations.

Step 10. Specify the dates of the next maintenance evaluation. Announce the dates of the next evaluation immediately. As necessary, identify any additional activities that will be evaluated. Establish new, higher performance targets for the next evaluation. Reinforce the policy of continuing evaluations.

QUALITY IS PROFITABILITY

Hand in hand with improving better maintenance, performance is the achievement of higher quality maintenance work. Common elements of performance and quality are evident in every aspect of equipment management. The maintenance program, for example, emphasizes the need of solid documentation to ensure that all of its elements are well understood and, as a result, carried out effectively. Similarly, quality assurance requires effective documentation of exactly how an organization is to ensure quality.

Total quality management (TQM) integrates the functions and processes within an organization to achieve continuous improvement of services and products to satisfy the customer. TQM targets the consumer as the customer who, in turn, makes the final decision on the quality of the product. But the plant and the corporation control whether the product is of sufficient quality to satisfy the consumer. If it is, they can compete and be profitable. If not, they go out of business. "One factor above all others, quality, drives market share. And when superior quality and large market share are both present, profitability is virtually guaranteed" (Buzzell and Gale 1987).

It is useful to examine the thinking of the quality experts in relating quality to management. They identified the key issues of TQM (Mercer 1991):

- The cost of quality is the measure of nonquality (not meeting customer requirements) and a measure of how the quality process is progressing.
- Total quality management is a cultural change that appreciates the primary need to meet customer requirements, implements a management philosophy that acknowledges this emphasis, encourages employee involvement, and embraces the ethic of continuous improvement.
- There must be enabling of the mechanisms of change, including training and education, communication, recognition, management behavior, teamwork, and customer satisfaction programs.
- Implementing TQM happens by defining the mission, identifying the output, identifying the customers, negotiating customer requirements, developing a "supplier specification" that details customer objectives, and determining the activities required to fulfill those objectives. Certainly, one key objective of equipment management is to align the maintenance program to bring quality service to operations.
- There must be management behavior that includes acting as role models, use of quality processes and tools, encouraging communication, sponsoring feedback activities, and fostering and providing a supporting environment. A supporting environment is provided by the managers' production strategy.

Downtime or lack of maintenance quality denies profitability. Just as the production strategy breaks away from antiquated maintenance practices, cultural change must precede TQM. Training and education are the means of changing culture along with management leadership. W. Edward Deming's (1982) universal fourteen points further emphasize this belief:

1. Create consistency of purpose with a plan.
2. Adopt the new philosophy of quality.
3. Cease dependence on mass inspection.
4. End the practice of choosing suppliers based solely on price.
5. Identify problems and work continuously to improve the system.
6. Adopt modern methods of training on the job.
7. Change the focus from production numbers (quantity) to quality.
8. Drive out fear.
9. Break down barriers between departments.
10. Stop requesting improved productivity without providing methods to achieve it.
11. Eliminate work standards that prescribe numerical quotas.
12. Remove barriers to pride of workmanship.
13. Institute vigorous education and retraining.
14. Create a structure in top management that will emphasize the preceding thirteen points every day.

Top management must emphasize cooperation. In doing so, they are providing the method to achieve productivity. Education and retraining are integral to the quality implementation process. Participative evaluations and solid information identify problems that inhibit implementation progress. J.M. Juran introduced the managerial dimensions of planning, organizing, and controlling. Juran focused on the responsibility of management to achieve quality and the need for setting goals. He believed that quality was the fitness for use in terms of design, conformance, availability, safety, and field use. He focused on top-down management and technical methods rather than worker pride and satisfaction (Juran 1951). Juran's ten steps to quality improvement are as follows:

1. Build awareness of opportunities to improve.
2. Set goals for improvement.
3. Organize to reach goals.
4. Provide training.
5. Carry out projects to solve problems.
6. Report progress.
7. Give recognition.
8. Communicate results.

9. Keep score.
10. Maintain momentum by making annual improvement part of the regular systems and processes of the company. (Juran 1951)

Juran's view of quality was a broader concept than Deming's focus on more technically oriented statistical process control. But it was Armand Feigenbaum who promoted integrating efforts to develop, maintain, and improve quality by the various groups in an organization.

Feigenbaum emphasized that quality control must be built in at an earlier stage of the quality process. Thus, his total quality control approach more closely resembles today's view than either Deming's or Juran's TQM (Feigenbaum 1990). Philip Crosby, author of *Quality Is Free*, got more direct attention of the industrial sector with the view that poor quality loses about 20% of revenues, most of which could be avoided by adopting good quality practices (Crosby 1979). He also had additional principles to be added to the growing list of points to be followed by those embarking on the implementation of TQM. Crosby added

- Quality is defined as conformance to requirements, not "goodness."
- The system for achieving quality is prevention, not appraisal.
- The performance standard is zero defects, not "that's close enough."
- The measurement of quality is the price of nonconformance, not indices.

Crosby stressed motivation and planning rather than statistical process control and the problem-solving techniques. He stated that "quality is free" because the small costs of prevention will always be lower than the costs of detection, correction, and failure. His own 15 points include the following:

1. Management commitment—Top management must become convinced of the need for quality and must clearly communicate this to the entire company by written policy, stating that each person is expected to perform according to the requirement or cause the requirement to be officially changed to what the company and the customers really need.
2. Quality improvement team—Form a team composed of department heads to oversee improvements in their departments and in the company as a whole.
3. Quality measurement—Establish measurements appropriate to every activity in order to identify areas in need of improvement.
4. Cost of quality—Estimate the costs of quality in order to identify areas where improvements would be profitable.
5. Quality awareness—Raise quality awareness among employees. They must understand the importance of product conformance and the costs of nonconformance.

6. Corrective action—Take corrective action as a result of steps 3 and 4.
7. Zero-defects planning—Form a committee to plan a program appropriate to the company and its culture.
8. Supervisor training—All levels of management must be trained in how to implement their part of the quality improvement program.
9. Zero-defects day—Schedule a day to signal to employees that the company has a new standard.
10. Goal setting—Individuals must establish improvement goals for themselves and their groups.
11. Error cause removal—Employees should be encouraged to inform management of any problems that prevent them from performing error-free work.
12. Recognition—Give public, nonfinancial appreciation to those who meet their quality goals or perform outstandingly.
13. Quality councils—Quality councils, composed of quality professionals and team chairpersons, should meet regularly to share experiences, problems, and ideas.
14. Do it all over again—Repeat steps 1 to 13 in order to emphasize the never-ending process of quality improvement. (Crosby 1979)

Despite some minor differences among these experts, a number of common themes arise:

- Inspection is never the answer to quality improvement, nor is "policing."
- Involvement of and leadership by top management are essential to the necessary culture of commitment to quality.
- A program for quality requires organization-wide efforts and long-term commitment, accompanied by the necessary investment in training.
- Quality is first and schedules are secondary. (Ross 1995)

The understanding of the process of implementing quality control has direct parallel to the successful maintenance strategies. Both processes begin with a firm commitment by management that a new direction has been established. Then very specific policies are provided to ensure that the enabling interdepartmental cooperation required takes place. Now education is provided to dispel the fears that might inhibit the new relationships that are required between departments. Evaluations are conducted to ensure the correct identification of critical problems. Then the specific tools that make each process possible are added: programs, systems, information, and technology. Finally, goals are established and monitoring is set in place to measure progress and ensure continuous advancement. Continuous improvement then becomes a measure of quality performance.

GETTING THERE WITH ISO 9000

ISO 9000 is a set of worldwide standards that establishes requirements for the management of quality. Unlike standards for products, these standards are guidelines, which, if followed, can help ensure that a company can establish a quality management system. In turn, a company with a certified quality management system can assure their customers that their products or services will meet quality levels imposed on them as a supplier. ISO standards help to ensure that a certified company has a quality system that will enable them to meet required quality standards. ISO 9000 consists of five international quality standards developed by the International Organization for Standardization (ISO) in Geneva, Switzerland. Collectively, the ISO standards are used by the 12-nation European Economic Community to provide a universal framework for quality assurance, primarily through a system of internal and external audits. Their utilization is quickly spreading worldwide to ensure universal application of quality in an increasingly global marketplace. The ISO standards are generic in that they apply to all functions and all industries, from banking to manufacturing (Johnson and Kantner 1992).

Although ISO 9000 is a road map to using the other standards in the series, ISO 9001, 9002, and 9003 standards are guidelines for establishing quality assurance programs for various types of industries or activities, specifically,

- ISO 9001 guides quality conformance by engineering firms, and construction and manufacturing companies during design, development, production, installation, and servicing.
- ISO 9002 guides quality assurance for manufacturing during production, processing, and installation.
- ISO 9003 specifies quality conformance in final inspection and testing by small shops or equipment distributors.
- ISO 9004 provides guidelines for developing and implementing a quality system.

IMPLEMENTING THE SYSTEM

ISO standards do not spell out the means of implementation. That is up to the company. However, after a decision has been made to adopt the standards and seek certification, the following steps can facilitate implementation where maintenance is concerned.

1. Get your plant management to provide quality objectives for maintenance within its operating strategy. For example, have them point out that preventive maintenance (PM) inspections, testing, and monitoring ensure reliable equipment that will consistently meet quality product or service requirements.

2. Ensure that plant management provides specific policies requiring the efficient conduct of preventive maintenance as well as support of maintenance by operations, for instance.
3. After determining the scope of production to be certified, list the production equipment involved, to the component level, necessary to produce the specified product quality.
4. Next, document the maintenance program that will be applied to this equipment. This documentation should include elements such as preventive maintenance, planning and scheduling, component replacement, project management, work control, material management, and how they interact. In addition, the duties of key personnel as they carry out these actions should be prescribed. Where appropriate, specific checklists for PM services or tasks lists for major work like overhauls are necessary. Finally, add definitions of work loads, terminology, and so forth, to clarify the documentation.
5. Finally, be able to identify the actions required to meet the required maintenance objectives. This means that you must establish an audit procedure that can determine how well the maintenance program, the organization that carries it out, and the information system that helps control actions help meet the reliability levels specified for equipment involved.

The role of suppliers or vendors to maintenance should not be overlooked. If, for example, a vendor rebuilds components that maintenance will install or overhauls units used in production, those suppliers must be included in the documentation.

REGISTRATION

Most industries such as chemicals, food, pharmaceutical, manufacturing, and process industries will seek registration of a quality system to conform with ISO 9002. Keep in mind that registration does not qualify a company for technical ability. Rather, if the capability to sustain a quality system has been defined, a judgment can be made wherein the company can adhere to procedures and policies that will ensure a quality product can be produced.

A criticism often aimed at ISO registration is that it is only a verification that good records are being kept on what might be a bad process. As some put it, a company could manufacture "life jackets made out of concrete" and be certified if it kept good records on its production process. The fact that the product failed to carry out its intended purpose is another matter, outside of the "quality" assurance system.

Three steps to the registration process are

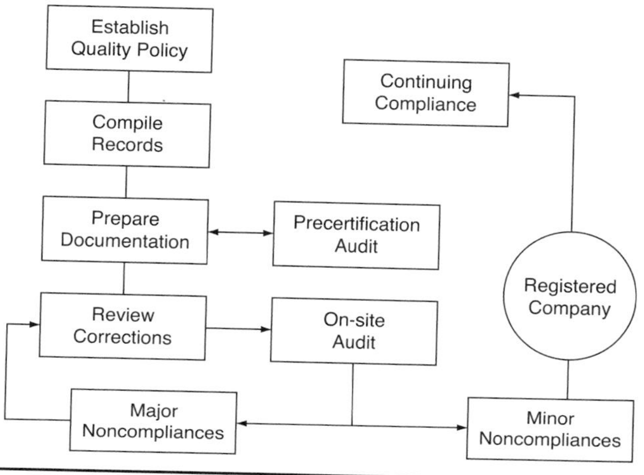

FIGURE 18-1 The registration process

1. Appraisal of maintenance documentation as it applies to meeting product reliability requirements;
2. Assessment of how well maintenance does what they said they were going to do; and
3. Presentation of findings and recommendations for corrective action.

After establishing company policies and procedures for implementing a quality system, necessary documentation is assembled (see Figure 18-1). Then a courtesy precertification audit is held to identify those aspects that need further work. Corrections are made and, if there are no major deficiencies, the registered company corrects the minor problems. After certification, periodic audits are requested or conducted to better ensure continued improvement.

Documentation of a maintenance program means that every aspect of requesting work, and thereafter classifying, planning, scheduling assigning, controlling, and measuring it, must be spelled out. Justification for such documentation is reflected in the axiom that "if you haven't written it out, you probably haven't thought it out." Registration can help solve many maintenance problems. A common shortcoming of most inadequate maintenance departments is a failure to spell out their programs. Consider the following situation:

> More than 45% of maintenance organizations attempting teams failed because the maintenance program was confusing and inconsistent. Team members (craftsmen) had only experience in diagnosis and repair. They knew little about "work control" techniques because previously, supervisors always told them what to do and when. Thus, they failed.

A poorly defined maintenance program is perceived differently by each individual. Thus, in the above-mentioned situation, it is unlikely that there could be effective maintenance, much less adequate quality assurance. Of the failures to successfully implement information systems, most point to poorly defined maintenance programs. Many do not appreciate that the new system is only the communications network to support the maintenance program, not the program itself. Thus, a prudent plant manager, struggling with the need to improve maintenance, might use ISO registration to better focus maintenance in satisfying program definition as well as ISO registration. ISO registration could elevate maintenance to the prominent level it must have in the company's operating strategy. Generally, the quality of the documentation should be good enough so that as people come and go, change jobs, and forget a procedure, documentation ensures continuity. Literally, the auditor perceives documentation to be adequate if the situation occurred where all personnel involved in maintenance were replaced, and new people of similar qualifications would be able to continue carrying on at the same effectiveness level by following the documentation. Therefore, the advice given to firms wishing to certify is "say what you do, then do what you say, and be able to prove it."

THE ASSESSMENT STEP

Many organizations are reporting that auditors are looking at maintenance but only as a sideline to the production process. Some have reported that they did nothing and their plants were certified nevertheless. If one were to speculate, this situation may soon change. ISO registration is spreading rapidly, and with that demand will come a more definitive assessment of maintenance practices.

Auditors will focus more on supporting activities like maintenance as registration demand increases. Auditors should be expected to ask

- Whether the maintenance program ensures that the equipment being maintained is able to reliably meet the process control requirements that will yield a quality product?
- If maintenance and operating personnel understand the program? Expect randomly selected personnel to be quizzed.
- Whether all personnel who will carry out the maintenance program are properly trained (including operators performing maintenance tasks) to successfully carry out prescribed services and repairs? An operator might be asked to demonstrate how to calibrate a control device, for example.
- If all procedures are up to date with the latest techniques? Is maintenance using only oil sampling when tribology techniques (the study of surfaces in relative motion) are called for?

- If an audit can ensure that a program is functioning properly, the organization is operating effectively, and the information is adequate to control work?

In short, today's auditors are learning the ropes, as evidenced by the number of seminars being offered to instruct them. Those aspiring to be auditors may now be focusing on the highlights, not the intimate details. With more demand by customers for authentic certification, the expectation is that maintenance will be more closely scrutinized in the future.

THE THIRD-PARTY AUDIT

The third-party audit is a prerequisite to certification. The traditional two-party quality audit relies on a buyer–seller relationship, in which the buyer (customer) "audits" the supplier. For example, a milk producer might audit the production process of the supplier that makes the paper used to manufacture its milk cartons. This is a healthy, direct application of quality assurance. A third-party audit is required because a customer with a hundred or more suppliers, each with their own specific requirements, would find it difficult to assess each supplier. From a customer's point of view, it is beneficial when all suppliers are judged by a single set of criteria. The third-party audit solves this problem by placing importance on systems that ensure quality. With financial audits, an impartial third-party audit is required. It makes sense to apply the same standard to quality systems. This is particularly important in helping to guarantee quality across international borders. Thus, the independent third-party registrar is able to certify that the quality system meets the requirements of ISO 9000 universally.

The third-party audit and subsequent certification, if achieved, should be viewed as a means to be achieved, not an end. The importance of preparation for certification lays not so much in the certification itself, but in the quality system that results from the effort leading to it. Notwithstanding, the preparation can result in a better plant as well. Now it can better be seen how the act of registration is a means whereby the plant manager can, concurrently, improve maintenance while preparing for certification. The customer is the ultimate beneficiary of the quality system. Any effort to obtain ISO 9000 certification without customer communication can be a waste of time and a compromise of any system that may result. Customers who are informed of the company's efforts to better supply them are impressed. They expect that such efforts will yield quality products whether one certifies or not. This is much like the effort made to try to learn a foreign language. Credit is awarded for effort, even though fluency is not achieved. Certification is a beginning, not an end. Ongoing evaluation, feedback, and fine-tuning must follow to ensure continuing certification. Thus, plant managers and maintenance must provide auditing procedures that will ensure continuing certification.

SUMMARY

Evaluations establish the current maintenance performance levels by identifying those activities needing improvement as well as those being performed well. The evaluation not only confirms performance but is the starting point for any improvement effort. The most important result of an effective evaluation is the education of personnel and their commitment to providing support for improving maintenance.

Guidelines to implement total quality management parallel those necessary to successfully implement a maintenance strategy. From firm management commitment and guidance to department interaction, a study of TQM is excellent preparation.

REFERENCES

Buzzell, R.D., and Gale, B.T. 1987. *The PIMS Principles: Linking Strategy to Performance.* New York: Free Press. p. 87.

Crosby, P. 1979. *Quality Is Free*. New York: McGraw-Hill.

Deming, W.E. 1982. *Quality, Productivity, and Competitive Position*. Cambridge, MA: Center for Advanced Engineering Study.

Feigenbaum, A.V. 1990. The criticality of quality and the need to measure it. *Financier* (October): 33– 36.

Johnson, P.L., and Kantner, R. 1992. *All About ISO 9000*. Southfield, MI: Perry Johnson.

Juran, J.M. 1951. *Quality Control Handbook*. New York: McGraw-Hill.

Mercer, D. 1991. Total quality management: Key quality issues. In *Global Perspectives on Total Quality*. New York: Conference Board. p. 11.

Ross, J.E. 1995. *Total Quality Management*, 2nd ed. Delray Beach, FL: St. Lucie Press. p. 8.

APPENDIX

Maintenance Performance Evaluation

The maintenance performance evaluation is made up of 14 sections. These are the type of standards against which maintenance should be evaluated.

Section 01–Working Environment: Assesses how effectively local managers have created a positive working environment so that maintenance will be effectively supported, efficiently used, and properly carried out.

Maintenance must be an integral part of a plant's operating strategy. Maintenance cannot be viewed as a "necessary evil." All plant departments must help maintenance to succeed. They depend on operations, for example, to make their equipment available when maintenance is scheduled. Within the operating strategy, purchasing, for example, must be accountable for delivering vital replacement components at the specified time. An effective operating strategy ensures that maintenance receives the cooperation and support essential for success. It also ensures that maintenance interacts accountably with other departments. Plant managers should specify responsibilities for the conduct and support of maintenance.

1. Management has provided a well-publicized operating strategy, specifying responsibilities of all departments in the conduct or support of maintenance.
2. Management has provided guidelines for each department in the development of their programs. As a result, there are well-documented programs for maintenance, as well as for operations, purchasing, warehousing, and so on.
3. Maintenance has documented their program effectively.
4. Maintenance has informed other departments how their program operates and how the departments can help to carry it out.
5. Each department has a well-defined program explaining how other departments interact with them. For example, warehousing has prescribed

a procedure for returning unused parts, and maintenance follows it carefully.

6. Management verifies the responsiveness of maintenance.
7. Management has created a positive working environment for maintenance.
8. Management has set specific schedule compliance goals for operations and acts promptly to ensure better compliance when goals are not met. For example, if preventive maintenance schedule compliance by operations was only 34% against a target of 85%, management would act promptly to determine the cause and correct it.
9. Management investigates instances of excessive downtime to determine the causes and correct them. For example, if the replacement of an engine were delayed because it was not delivered to maintenance on time by purchasing, management would act promptly to determine the cause and correct it.
10. Cost and performance information required to assess overall maintenance performance is used effectively by management to identify and correct significant maintenance problems.

Section 02–Operations Support: Assesses how well operations utilizes maintenance services.

A perfect maintenance program could fail if operations does not make equipment available after maintenance has organized their resources and obtained approval of the schedule. However, with pressure to meet production targets, operations might give higher priority to keep using equipment than getting essential maintenance done. Compounding this problem are daily requirements to explain to local managers why production targets are not met. Then if vital equipment fails and production targets are not met, maintenance could become the scapegoat. This scenario is commonplace among those operations struggling for profitability. The realistic steps that must be taken to ensure this does not happen start with local managers making operations responsible for the "fact" of maintenance. Maintenance has a program and is prepared to carry it out. Operations must use maintenance services effectively. In turn, operations personnel must understand the maintenance program and align their operating procedures with it.

1. Operations personnel have a good knowledge of the maintenance requirements of the equipment they operate.
2. Operations personnel are familiar with pertinent aspects of the maintenance program.
3. Equipment is operated carefully and correctly to avoid unnecessary damage.

4. Operators report problems promptly.
5. Operators perform maintenance-related actions such as cleaning, adjusting, and servicing of equipment effectively.
6. Operations personnel help to identify the timing and cost of future maintenance work such as overhauls.
7. Operations requires that preventive maintenance services be completed on time.
8. Operations is concerned with job cost and work quality.
9. Operations does not overload maintenance with unexpected work or unreasonable demands.
10. Work requested by operations is necessary, feasible, and correctly funded.
11. Operations makes equipment available for maintenance according to an approved major maintenance (weekly or shutdown) schedule.
12. Operations complies with scheduled preventive maintenance services by bringing mobile equipment to the garage or shutting down fixed equipment promptly at the time agreed upon in the schedule.
13. Maintenance and operations cooperation is evident at every level of their respective organizations.

Section 03–Staff Support: Assesses how efficiently staff departments such as warehousing, purchasing, or accounting support maintenance.

Maintenance depends on reliable service by staff departments such as warehousing or purchasing. Without the right materials in the right quantity at the time needed, a quality maintenance program could fail.

Similarly, if maintenance wishes to have first-rate, quality service from staff departments such as warehousing or purchasing, they must comply with procedures established by these organizations. Maintenance should also, for example, provide sufficient lead time, as in obtaining direct charge (purchased) materials such as engines or drive motors through purchasing. Abuses of material control procedures by maintenance should be avoided. For example, if 6 items are needed for a job, don't draw 12 and stuff the unused 6 in a drawer. Similarly, maintenance should follow field-reporting procedures for accounting. If, for instance, accounting requires that man-hours used be reported by the craftsmen and verified by supervisors, the procedure should be followed to ensure accurate data.

1. Staff personnel understand the maintenance program.
2. Maintenance has made a genuine effort to ensure that staff departments have been fully informed of the details of the maintenance program so that they can support it effectively.

3. Staff personnel are service oriented.
4. Recommendations for better services by staff departments are acted on promptly.

Section 04–Maintenance Program: Assesses how well the maintenance program has been defined and the plant informed how they can help.

The maintenance program should prescribe what maintenance does, how they do it, who does what, and why. Because the maintenance program includes coordination with operations or obtaining services from warehousing, for example, it must clearly describe these essential interactions. The absence of a well-defined maintenance program or the ineffective communication of program elements to other departments penalizes maintenance as it denies them needed support. Typically, a well-defined and effective maintenance program spells out the interaction of all departments as they request or identify work, classify it to determine the best reaction, plan selected work to ensure it is accomplished efficiently, and schedule the work to ensure it is performed at the best time with the most effective use of resources.

In addition, the maintenance program specifies how work is assigned to personnel in a way that ensures each person has a full shift of bona fide work. Then as work is performed, the program establishes work control procedures to ensure quality work that is completed on time. In addition, the program specifies how completed work is measured to ensure timely completion, under budget, with quality results. The maintenance program should also prescribe a means of periodic evaluations to identify and prioritize improvement needs.

1. Maintenance has a well-defined program.
2. The program prescribes effective ways to identify new work using preventive or predictive maintenance services, analysis of repair history, costs, and so on.
3. Clear procedures are prescribed for requesting new work.
4. The maintenance program specifies how nonmaintenance work such as construction, new equipment installation, or major equipment modifications will be submitted, assessed, funded, and carried out.
5. The program has defined what "maintenance" consists of to eliminate any confusion with other work it might do, such as construction.
6. As part of its program definition, maintenance has defined everyday terminology. It has, for example, explained the difference between a rebuild and an overhaul, or between corrective maintenance and modification.
7. Within the program, maintenance has carefully defined the work load. Work load elements such as emergency repairs, preventive maintenance

services, or planned/scheduled maintenance, for example, have been defined, agreed upon, and published.

8. Criteria are in place for determining which work will be planned and scheduled. Criteria ensure the most effective use of planners and, concurrently, define work that is the sole responsibility of maintenance field personnel.
9. Procedures for scheduling are prescribed.
10. The maintenance program prescribes how work will be assigned to maintenance personnel and states clear objectives. A typical objective might be, for example, to ensure that each maintenance person has a full shift of realistic, bona fide, ready-to-be done jobs assigned to him or her.
11. Information required to carry out the maintenance program and control work properly is prescribed by the maintenance program. For example, the use of repair history or cost reports is prescribed to help identify new work. Similarly, the program might specify how the backlog, for example, is used to help determine whether maintenance is keeping up with the generation of new work.
12. Methods of controlling ongoing work are prescribed by the maintenance program. For example, on-site work control by supervisors, team leaders, leadmen, or others is prescribed. In addition, field reporting procedures are prescribed for personnel working on their own or working in a team environment.
13. The program prescribes how completed work will be measured to ensure a timely finish, within budget, against standards with quality results.
14. The maintenance program includes procedures whereby maintenance performance can be evaluated to identify improvement needs, prioritize them, and carry out improvement actions.
15. The maintenance program has been explained to all plant personnel to enable them to support it effectively.
16. The maintenance program has been sufficiently well-documented so that a person of equal qualifications to any incumbent can read the program details and deliver performance equal to the incumbent.
17. Implementation of maintenance strategies such as total productive maintenance or reliability centered maintenance is preceded by verification that maintenance fundamentals are mastered and use of information are effective.

Section 05–Maintenance Organization: Assesses how well maintenance has been organized.

Effective program execution requires a responsive, well-led organization that communicates well internally and externally. Different types of organizations are used to support different types of operations. First-line supervisors would be used when direct control of work is required versus teams empowered to work alone, for example. Organizations should satisfy principles of organization such as a clear division of work responsibilities. Leaders should be carefully chosen and criteria established for progression to higher leadership assignments. Pertinent training should be provided to develop and improve skills among craftsmen as well as their leaders. Periodic measurements should be made to evaluate and improve performance.

1. The type of maintenance organization best matches the needs of the operation being supported.
2. Organizational charts are complete and correct.
3. Job descriptions describe actual duties and qualifications to meet them.
4. The organizational structure facilitates efficiency and communications.
5. Criteria exist to ensure that the best candidates are chosen as leaders.
6. Requirements for promotion are clearly stated.
7. Leaders have specific responsibilities and sufficient authority to carry them out.
8. Leaders are rated on their performance, advised of results, and counseled as required.
9. Incompetent leaders are promptly replaced.
10. Crafts personnel seek leadership positions and seldom return to their tools.
11. Personnel are empowered to make decisions on actions they control.
12. Selection of personnel for senior maintenance leadership positions includes personnel from other departments, providing they meet the selection criteria.
13. During peak periods of absenteeism, personnel still get work done.
14. Leaders have adequate time and opportunity to carry out their key duties.
15. Maintenance personnel are well-informed on the details of their program.
16. There is clear evidence of a cooperative working relationship within maintenance.
17. Maintenance works harmoniously with other departments.
18. Training to improve craft skill levels is provided.

19. Skill levels of craft personnel are clearly established and verified.
20. Effective supervisory training is provided.
21. Organizations are evaluated regularly to identify improvement opportunities.

Section 06–Information: Assesses how effectively maintenance uses information.

Information is the basis for decisions on the control of work, the use of resources, and management of cost. It must be complete, accurate, and timely. It should be made available to the right personnel and presented in a useful format. The work order system, an element of the information system, provides a means of requesting new work and converting it into jobs to be planned, scheduled, assigned, and controlled. As work is being done, the work order system brings field data into the information system where it is converted into information. Because the information system is the communications network for department interaction as the maintenance program is carried out, it must be able to be used competently by all personnel. Thus, the system must be easy to use and provide the information necessary to manage all aspects of maintenance.

Regarding the backlog, a listing of all open work orders is not a "backlog." The backlog measures how well maintenance keeps up with the generation of new work, and its data helps to determine the proper size and craft composition of the work force. The backlog lists only work that has been planned and scheduled for which man-hours, by craft, have been estimated. Therefore, the backlog is measured in estimated man-hours by craft, not by the number of work orders. Thus, emergency repairs, unscheduled repairs, and routine, repetitive preventive maintenance services are not in the backlog. They are, however, listed in the open work order list.

1. Field data such as man-hours used or parts consumed are reported properly.
2. Craftsmen report field data such as man-hours used, and supervisors verify it.
3. The work order system covers all types of work.
4. Work order procedures are simple and easy to use. For example, every craftsman and equipment operator can request work efficiently and quickly.
5. Small jobs are entered into the work order system in the easiest and simplest way possible.
6. The work order system provides for easy control of routine, repetitive activities such as shop cleanup, training, or buildings and grounds work. For example, standing work orders are used for routine, repetitive

functions like shop cleanup or to represent groups of low-maintenance-cost equipment.

7. The work order system lends itself to easy job planning and resource estimating.
8. The work order system provides an effective priority-setting scheme.
9. Work order approval is required for selected types of high-cost work.
10. When verbal orders are used to request work, the system precludes loss of control of the work or loss of information about it.
11. Meaningful indices are provided to enable managers to assess overall maintenance performance.
12. Operations reviews maintenance cost and performance.
13. All who need information can get it.
14. Recipients know how to use information.
15. Actual and estimated man-hours are compared on major jobs.
16. Information shows labor use. For example, man-hours used on preventive maintenance versus emergency repairs or scheduled maintenance are provided.
17. Information is provided on absenteeism.
18. Overtime use information is available.
19. The backlog shows whether maintenance is keeping up with new work.
20. Man-hour estimates from the backlog help determine whether the work force is of sufficient size and composition by craft to carry out the work load.
21. Backlog information is of sufficient reliability that management will commit to changing the work force size if backlog data so indicates.
22. An administrative open work order file or list shows all work orders open or being worked on. (This is different from the backlog.)
23. Cost and performance on major jobs are available.
24. Costs are available for individual units of equipment.
25. Costs are provided for individual major components (e.g., drive motors) on units of production equipment.
26. Costs are available on functions such as grass cutting.
27. Cost information is summarized by cost centers.
28. Actual costs and budgeted costs are compared month and year to date by cost centers.
29. Repair history traces significant equipment repairs and failure patterns.

30. Repair history includes data on the life-span of critical components.
31. Report or display formats are clear and easy to read.
32. The information system provides all information required for control of work as well as cost and performance management.

Section 07–Preventive Maintenance and Technology: Assesses preventive maintenance and the use of modern technology.

The focal point of maintenance technology is the preventive maintenance (PM) portion of the maintenance program. Preventive maintenance should successfully extend equipment life with services such as cleaning, adjustment, lubrication, and minor component replacements. It should also help avoid premature failures through timely inspection, condition monitoring, and testing. Successful conduct of PM services should result in fewer emergency repairs, and more work should be able to be planned. Planned work is better organized and ensures the best use of labor. Planned work is performed more productively and deliberately, yielding higher-quality work. PM services should not include individual major jobs such as major component replacements or overhauls. Such jobs are planned and scheduled and occur at irregular intervals. PM services, on the other hand, are carried out at regular or variable intervals. This distinction makes it possible to accurately estimate the total man-hours required to carry out PM services. If overhauls, for example, were lumped under PM services, the labor required for preventive maintenance could not be determined. Each PM service is planned at the time the service was originally established. Thereafter, PM services are repetitively scheduled according to prescribed service intervals. Therefore, when the percentage of planned work is calculated, it should not include repetitively scheduled PM services.

The conduct of maintenance engineering ensures the correct application of modern technology to maintenance for verifying the maintainability and reliability of equipment. Specifically, maintenance engineering will ensure that PM services are correct and include all applicable predictive techniques. Maintenance engineering will also develop standards for periodic maintenance tasks such as major component replacements. Standards provide targets against which maintenance performance can be assessed. Quality standards, for example, prescribe how the job should be done and what the final product should be. They typically include task lists, bills of materials, and tool lists. Quantity standards prescribe the amount of labor by craft, job duration, and so forth. Maintenance engineering also establishes commissioning procedures for new equipment, assesses equipment modifications, and verifies the adequacy of nonmaintenance project work such as new equipment installations or modifications, whether performed by contractors or work force personnel. Maintenance engineering tasks may be carried out by a dedicated staff or by supervisors and craft personnel. The increasing complexity of equipment and the greater need for reliability requires that the types of services provided by maintenance engineering be carried out consistently and with the highest level of technical expertise. A contractor might provide certain technical services such as vibration analysis or infrared testing. However, all such services should be part of the overall PM effort. They should be coordinated by maintenance personnel, and the quality and effectiveness of contractor services verified.

1. Preventive maintenance services are well-defined. That is, each service has a carefully prepared checklist, service times are prescribed, service frequencies are designated, and a method is provided to measure schedule compliance and overall effectiveness.
2. All maintenance personnel and operating departments understand preventive maintenance. They know, for example, what constitutes preventive maintenance services, and how they are carried out and controlled.
3. Management understands and strongly supports preventive maintenance. As evidence of this support, for example, they have set schedule compliance targets and have made operations accountable for making equipment available for services.
4. Preventive maintenance services are oriented toward the detection of equipment deficiencies and they successfully uncover them well in advance of potential equipment failures.
5. Man-hours required, by craft, for the conduct of each preventive maintenance service has been established.
6. Completion of scheduled preventive maintenance services are verified, including a report of schedule compliance to management.
7. When new equipment is added or existing equipment modified, changes are promptly made to applicable preventive maintenance services.
8. All preventive maintenance services are reviewed periodically to determine their adequacy.
9. Preventive maintenance services are carried out with care and diligence by craft personnel.
10. Where appropriate, operations personnel perform preventive maintenance–related tasks completely, efficiently, and correctly.
11. Key maintenance personnel such as supervisors or team leaders follow up to ensure that preventive maintenance services are being done on time.
12. Operating personnel cooperate by making equipment available for services.
13. Condition-monitoring services utilize predictive techniques skillfully.

14. Where specialized condition-monitoring devices such as on-board computers are in use, maintenance personnel can effectively download data, analyze it, and utilize the resulting information effectively.
15. Extensive repairs are not carried out until preventive maintenance services are completed. This avoids missing more serious deficiencies by running out of time before inspection, testing, or servicing is completed.
16. Preventive maintenance services for fixed equipment are linked together in routes to avoid unnecessary travel (backtracking) to visit equipment along the route.
17. Within the prescribed service intervals, preventive maintenance services for mobile equipment are jointly scheduled with operations to avoid unnecessary interruption of production.
18. The manner in which preventive maintenance services have been carried out has contributed to better overall maintenance performance. This is confirmed, for example, with fewer emergencies, more planning, and less downtime.
19. Equipment maintainability, reliability, and performance are assessed regularly.
20. Repair history is analyzed to identify chronic, repetitive equipment problems.
21. Field investigations are used to confirm equipment problems identified initially by information such as repair history.
22. New equipment is fully maintainable before being put into service.
23. Repair techniques are prescribed and training provided.
24. A modern technical reference library exists.
25. Standards are used for jobs on which work quality, cost, and performance are important.

Section 08–Planning and Scheduling: Assesses the effectiveness of planning and scheduling.

Planning ensures that jobs can be completed more efficiently. It determines labor, materials, tools, and supporting equipment needs in advance to ensure their availability for specific jobs. Planned jobs use less labor and are completed with less downtime because personnel work more efficiently. Because planned work is carried out more deliberately, the work is done more carefully, resulting in better work quality.

Criteria should be in place for determining which jobs are to be planned. There must be an effective procedure for planning. Planners should be used primarily for planning work. Maintenance and operations schedule major work jointly to establish the best timing for a job and to avoid interruptions to production while utilizing maintenance resources most efficiently. Repetitive tasks such as PM services are scheduled routinely. Repair history provides information on the life-span of major components so that the interval of time before the next replacement can be estimated among all similar components. This is often called the mean (average) time before failure, or MTBF. For example, a specific type of gear case operating under particular conditions might demonstrate a life-span of 26,000 operating hours. This information enables maintenance to "forecast" the next repetition of the event, plan it, and notify purchasing when it is time to procure the necessary components. Similarly, the forecast alerts operations of upcoming major work.

1. Criteria specify the type of work to be planned.
2. There is a clear division of work responsibility between planners and field personnel. For example, planners plan major jobs for execution by crews. Alternately, crews are responsible for organizing and carrying out all jobs, such as unscheduled and emergency repairs, that do not meet the criteria for being planned.
3. Planning procedures are complete and clear.
4. Work order documents facilitate the planning task.
5. Cost and performance information is available on planned jobs from inception to completion.
6. A forecast identifies most major future jobs to be planned and scheduled. A typical job might be the replacement of a major component such as an engine or drive motor.
7. Each forecasted major job has standards, including a task list explaining what to do, how to do it, and the sequence; a bill of materials keyed to the task list; and a coordinated tool list.
8. The use of standards has simplified the planning task and allowed individual planners to plan more jobs.
9. Preventive maintenance inspections, testing, and condition monitoring are major sources of planned work.
10. Crews prefer work that is planned.
11. When appropriate, nonmaintenance work like construction is planned.
12. Planning has reduced downtime required to complete major jobs.
13. Operations prefers maximum planned work and supports efforts to achieve it.
14. Criteria exist for the selection of planners.
15. Planners have ample time to plan.
16. There are enough well-trained planners.
17. The percentage of planned work is increasing.

18. Maximum planned work is possible in the plant operating environment.
19. Management policies guide maintenance and operations in developing and carrying out effective scheduling procedures.
20. There are regular operations–maintenance scheduling meetings that ensure work is scheduled when it interferes least with operations and makes the best use of maintenance resources.
21. Scheduling procedures are well-explained and effective.
22. All planned work is scheduled with operations.
23. Operations approves the weekly schedule before work is done.
24. Management verifies that operations makes equipment available and maintenance performs work according to the approved schedule.
25. Nonmaintenance work, like construction, is included in the weekly schedule only after verification that it will not detract from doing essential maintenance.
26. Labor for scheduled work is allocated by priority.
27. There are regular operations–maintenance coordination meetings to adjust the weekly schedule in the event of unexpected delays.
28. Major maintenance is planned, scheduled, and completed within a reasonable period of time. This ensures that major jobs in the backlog are moved onto the active schedule and completed promptly to avoid further deterioration to the equipment.
29. Warehouse material shortages seldom interfere with carrying out the schedule.
30. Late or incorrect delivery of purchased materials seldom interferes with carrying out the schedule.
31. Management seldom overrides an approved schedule.
32. Supervisors prepare daily work plans for their crews based on day-by-day requirements set forth in the weekly schedule.
33. Crews or teams add small jobs to the daily work plan after noting that downtime opportunities have been created for specific units by the weekly schedule.
34. Crews or teams are alert to downtime opportunities created by emergency jobs. Materials, for example, might not be readily available for a specific job, but the equipment is available for other jobs to be performed on it.
35. Effective scheduling has increased equipment availability.

Section 09–Material Control: Assesses the effectiveness of warehousing and purchasing in supporting maintenance.

Having the right materials in proper quantity at the right time is essential to the success of maintenance. Therefore, procedures to identify, stock, purchase, or manufacture materials in proper quantities to make them available to maintenance as requested must be effective. But maintenance must also be able to determine what they need and specify the quantity and the time when they are required. The work order system, material control, and information system should jointly permit maintenance to obtain information on material costs and usage. Maintenance must follow procedures for obtaining material support.

1. Material control procedures are well-understood and effective.
2. Critical spare parts lists exist for equipment.
3. Parts interchangeability is well-documented.
4. Maintenance advises of items to be stocked and proper quantities.
5. Out-of-stock materials are replenished promptly.
6. The stock room is properly staffed with qualified personnel.
7. All maintenance personnel follow material issue procedures.
8. All maintenance personnel can identify stock materials.
9. Maintenance provides standard bills of materials for repetitive major jobs.
10. Stock issues are made to craftsmen on presentation of approved work orders.
11. Material cost information is available for units and components.
12. Purchasing and warehousing are well-informed of future material needs by maintenance.
13. Purchasing is given sufficient lead time to obtain materials.
14. Purchase order status can be tracked.
15. Drawings or data for the manufacture of spare parts are available.
16. On-site material delivery service is provided.
17. There are no unauthorized material storage areas.
18. Material control personnel effectively perform tasks such as refilling steel racks in shops or preparing kits of standards parts needed for repetitive preventive maintenance services.
19. The purchasing agent or warehouse manager demonstrates interest in maintenance activities with actions such as attending scheduling meetings or seeking information on forecasted future actions requiring material support.
20. Overall, maintenance personnel consider material control to be excellent.

21. Maintenance cooperation in using material control services is considered excellent by warehousing and purchasing personnel.

Section 10–Mobile Equipment, Facilities, and Services: Assesses how well garages, facilities, shops, and services conduct or support maintenance.

Shop facilities such as a machine, fabrication, or welding shop provide effective support for field maintenance work as well as for capital projects like new equipment installation. Mobile equipment maintenance can represent a total program, as in an open pit mine. Standards 1–16, listed below, apply to operations that are fully supported by mobile equipment, as in an open pit mine or underground operation. Standards 17–28 apply to facilities and services.

Mobile equipment standards:

1. There is an organized garage area for units awaiting service or repair.
2. Units are washed before work or services are performed.
3. Preventive maintenance services are carried out in properly equipped areas.
4. Equipment inspection deficiencies are reviewed to determine further repair actions before such repairs are initiated.
5. Repairs are not initiated until preventive maintenance services are completed. This ensures that all deficiencies are identified.
6. The garage is divided into efficient working areas.
7. Areas are designated for specific activities like servicing or long-term repairs such as overhauls.
8. The garage is organized to maintain all types of equipment.
9. Major maintenance for mobile equipment is well-planned.
10. Preventive maintenance services and major repairs are scheduled conveniently for operations.
11. Operations complies with the approved schedule.
12. Support facilities, such as a warehouse or tire shop, are conveniently located.
13. Quality control like road-testing is carried out.
14. Completed work is picked up promptly by operations.
15. Field repairs done by garage personnel are carried out efficiently.
16. Overall maintenance of mobile equipment is well-done.

Shop standards:

17. Shop activities are arranged to ensure an efficient flow of work.
18. Information is provided on the status of shop work.
19. Shop drawings or assembly data are complete and well-organized.

20. A workable, realistic priority system exists.
21. Shop personnel perform quality work.
22. Shop work is completed on time.
23. Field services such as transportation or rigging are well-coordinated.
24. Support personnel operate equipment efficiently.
25. Costs for support services are available.
26. There is sufficient equipment of the right type to support field work.
27. Debris is collected and disposed of promptly.
28. Roads are well-maintained.

Section 11–Nonmaintenance: Assesses how effectively nonmaintenance activities such as construction, new equipment installation, or equipment modification are evaluated, approved, engineered, funded, and carried out.

Maintenance often performs nonmaintenance work (capital projects) such as equipment installation or modification. Nonmaintenance work creates a new asset, as in the installation of a new conveyor. This differs from maintenance, which is the repair and upkeep of existing equipment, as in the replacing of a worn drive shaft. Also, maintenance is an operating expense, whereas nonmaintenance work is often capitalized. Nonmaintenance work should be reviewed to determine whether it is necessary and feasible as well as properly engineered and funded before it is assigned to maintenance. Policies must exist to ensure that nonmaintenance work does not interfere with essential maintenance. Policies should specify how and when contractors would be used.

1. Performing nonmaintenance work (e.g., construction projects) does not interfere with the conduct of basic maintenance.
2. A priority scheme allocates labor for maintenance versus project work.
3. The work order system contains elements allowing nonmaintenance work to be controlled effectively, even if the work is accomplished by a contractor.
4. Information is provided on the cost and performance of projects.
5. Equipment modifications are reviewed and approved before being made.
6. Completed equipment modifications are well-documented.
7. Nonmaintenance projects comply with approval and funding procedures.
8. Maintenance is provided with full instructions and engineering details of nonmaintenance projects it is to carry out.
9. The adequacy of completed nonmaintenance work, especially new installations, is verified to ensure maintainability regardless of who does the work.

Section 12–Performance and Improvement: Assesses short-term performance, such as timely work completion, schedule compliance, or availability. Examines long-term improvement efforts like productivity improvement or cost-reduction.

The primary means by which maintenance controls the cost of the work is the efficiency with which it installs materials. This makes effective use of labor essential. In turn, effective control of labor requires good planning. Therefore, maintenance must ensure that most work is planned. Maintenance must also be able to measure the work load and convert it into a work force of the correct size and craft composition. Maintenance should measure productivity to eliminate delays. These steps result in well-maintained, reliable equipment. Periodic evaluations should be carried out to identify essential improvement needs and prioritize them. Evaluation results provide the basis for developing an improvement plan. Regular evaluations should be an essential part of the business plan or operating strategy.

1. Management policies emphasize the control of labor.
2. Absenteeism, including vacations, is well-controlled.
3. Overtime is assigned properly, used effectively, and well-controlled.
4. Labor used on planned jobs is accurately estimated.
5. Labor is allocated for scheduled work with meaningful priorities.
6. Labor use is carefully reported and analyzed to ensure effective utilization.
7. Key personnel, such as supervisors or team leaders, are held accountable for the effective use of labor.
8. Work is assigned so that each crew member has a full shift of realistic, ready-to-be-performed work.
9. Supervisors control work efficiently at work sites or, alternately, by obtaining status reports from crews.
10. The amount and timing of nonmaintenance project work, such as equipment installation, is regulated to ensure that it does not interfere with the conduct of basic maintenance work.
11. Management policies specify regular productivity measurements. Improving worker productivity is a prominent goal of the plant.
12. Productivity is measured regularly.
13. Causes of delay are analyzed to reduce them and improve productivity.
14. There are regular evaluations that identify improvement needs, prioritize them, and form the basis for implementing them.

Section 13–Cost control: Assesses cost-control effectiveness.

Maintenance costs are forecasted against production targets by time periods: monthly, annually, and so on. This budgeting process establishes expectations against which actual expenditures are compared. With such a yardstick, performance is measured by comparing man-hours used, labor cost accrued, material cost accumulated, and total costs committed. The focus of these costs is equipment, buildings or facilities maintained, activities performed, or jobs completed. Maintenance controls costs primarily by the efficiency with which it installs materials (effective use of labor). But it can also influence costs by encouraging production to operate equipment correctly.

Overall cost-control effort should include adequate preventive maintenance, emphasis on planned work, and the use of cost-related information to help anticipate the need to prepare for and plan major work. Important opportunities for effective cost control are available in the efficient use of contractors, the reduction of the need to rework, and the management of warranties. In addition, attention to health, safety, environmental, and regulatory aspects can result in significant cost avoidance.

1. The maintenance budget is based on production targets and their relationship to anticipated maintenance events and their costs.
2. Capital expenditures are budgeted separately.
3. A cost summary compares actual and budgeted costs, on a monthly and year-to-date basis.
4. Maintenance costs are available for units of equipment and components.
5. Overhauls are subjected to cost and performance evaluation.
6. Supervisors and team leaders can explain excessive costs.
7. Supervisors and team leaders have cost-related performance targets such as tool loss reduction, accident prevention, or productivity improvements.
8. There are regular maintenance–production cost accountability meetings.
9. Maintenance has improved productivity.
10. Maintenance reviews warehouse stock to check consumption rates and identify obsolete parts.
11. Maintenance anticipates purchasing needs to ensure timely delivery and avoid extra costs like air freight.
12. Contractor use is justified by determining how effectively they can be used and the impact they will have on cost reduction.
13. Contractor work, whether on site or at distant shops, is effectively monitored for work quality, timely completion, and cost compliance.
14. Unnecessary "rework" is avoided.

15. Maximum advantage is taken to gain warranty payments on equipment, components, and parts.
16. Controls exist to emphasize, ensure compliance, and correct possible violations of health, safety, environmental, and regulatory requirements.
17. Decisions to purchase new equipment include consideration of the reliability, operability, maintainability, life-cycle costing, and productive capacity.
18. Capital productivity is considered by reducing operating costs to yield greater profitability and more effective use of capital investment.
19. There is an effective overall cost-control effort.

Section 14–General: Assesses general considerations to provide an overview of maintenance performance.

Rate the general conditions described below, considering all aspects of maintenance that you are personally familiar with.

1. Responsiveness of maintenance.
2. Quality of maintenance work.
3. Cooperation with other departments.
4. Effectiveness of maintenance managers.
5. Effectiveness of maintenance supervisors.
6. Competency of planners.
7. Competency of craftsmen.
8. Quality of preventive maintenance services.
9. Adequacy of the work order system (easy to use).
10. Quality of information (tells the whole story easily and accurately).
11. Adequacy of repair history (gives a complete picture).
12. Internal maintenance cooperation (harmony).
13. Effectiveness of planning and scheduling (they do what they say they will do and get it done on time).
14. Productivity of workers (good and getting better).
15. Cooperation by operations in making equipment available for maintenance.
16. Level of service provided by staff departments like the warehouse (first-rate).
17. Management interest in quality maintenance (dedicated and concerned).
18. Success in implementing programs such as people empowerment or introducing new technologies.

19. Success in implementing maintenance teams (productivity and pride up).
20. Success in implementing interdepartmental teams (operators and craftsmen appear as one).
21. Success in the involvement of personnel in beneficial change (look forward to next challenge).
22. Success in correcting long-standing problems like unnecessary delays that lead to wasteful labor practices (identify problems; act swiftly to correct them).
23. An improved safety record.
24. Good housekeeping practices have resulted in an orderly, attractive plant.

Glossary

adjustments – Minor tune-up actions requiring hand tools, no parts, and less than one half hour of time.

administrative information – Information used to communicate within maintenance and operate the maintenance information system.

area organization – A type of maintenance organization in which one supervisor is responsible for all maintenance within a reasonably sized geographical plant area.

autonomous maintenance – Performance of maintenance-related activities such as cleaning, adjustment, lubrication, minor repairs, or simple machine calibration by equipment operators. A cornerstone of total productive maintenance.

backlog – The total number of estimated man-hours, by craft, required to perform all identified, but incomplete, planned and scheduled work.

benchmarking – The systematic process of searching for best practices, innovative ideas, and highly effective procedures that lead to superior performance.

capital-funded – The funding authorization for nonmaintenance project work such as construction or new equipment installation.

capitalized – Funding for work that expands the plant operating capacity; gains economic advantage; replaces worn, damaged, or obsolete equipment; satisfies a safety requirement; or meets a basic need.

category of work – The types of work that make up the work load performed by maintenance: preventive maintenance, emergency repairs, and so forth.

component – A subelement of a unit of equipment such as the belt of a conveyor, the motor of a crusher, or the engine of a truck.

concept – The means by which a major program, such as maintenance, is carried out in relationship to its objective and the other programs that it supports (operations) or on which it depends (purchasing).

condition-based – Actions taken based on the condition of the equipment.

condition monitoring – The application of predictive maintenance techniques on a continuous, periodic, or on-demand basis to determine current equipment condition.

coordination – Daily adjustment of maintenance actions to achieve the best short-term use of resources or to accommodate changes in operation needs.

cost center – A department or area in which equipment operates or in which functions are carried out.

culture (of maintenance) – The prevailing knowledge, beliefs, and behavior of a specific group of people (e.g., maintenance). The view of others outside the group that describes their understanding of the others' knowledge, beliefs, and behavior. A stereotyping of a group.

decision-making information – Information necessary to control day-to-day maintenance and determine current and long-term cost and performance trends for management decisions.

deferred maintenance – Maintenance that can be postponed to some future date without further deterioration of equipment.

downsizing – The conversion of a larger organization into a smaller one while taking necessary actions to preserve the efficiency and effectiveness of the organization.

downtime – A time period during which equipment cannot be operated to perform its intended function.

empowerment – To allow employees a definitive role in the control of their activities.

engineering work order (EWO) – A control document authorizing use of the maintenance work force or a contractor for engineering project work such as construction.

equipment life cycle – An equipment life cycle includes selection, purchasing, commissioning, testing, operating, maintaining, overhauling, modifying, and replacing equipment.

equipment management strategy – A fully-coordinated, mutually supporting effort of every plant department and individual to achieve maximum reliability and productive capacity of critical equipment throughout its entire life cycle.

expensed – Maintenance work charged to the operating budget.

failure analysis – The study of equipment failure data and related field experiences to determine the source of chronic, repetitive equipment problems and the determination of actions to reduce or eliminate them.

failure coding – An indexing of the causes of equipment failure on which corrective actions can be based.

failure modes and effects analysis – A procedure that studies failure causes and ranks their risk, then applies the best technology to reduce occurrences while improving detection capabilities.

forecasting – A projection of anticipated major tasks that are predictable based on historical data.

function – An activity carried out on a unit or performed within a cost center, such as "power sweeping in department 06."

guidelines – General guidance provided by management as maintenance or other departments carry out their day-to-day functions. See also *policies*.

inspection – The checking of equipment to determine repair needs and their urgency.

ISO 9000 – A set of quality assurance standards that can be applied to any organization regardless of size or type. Used to develop a common approach to obtaining quality service or product.

level of service – The degree of maintenance performed to meet desired levels of equipment performance. A high level ensures little chance of failure, whereas a low level meets minimum requirements risking breakdowns on less critical equipment.

life-cycle costing – The cost incurred during the life-span of equipment to keep it in optimum operating condition.

maintainability – The degree to which equipment is able to be maintained effectively. Factors influencing maintainability might include simplicity, ease of access to its critical components, minimum degree of difficulty in replacing components or ease of carrying out common activities like lubrication. Other factors include alarms or signals alerting personnel to problems that, when corrected early, simplify repairs by making them less serious.

maintenance engineering – The use of engineering techniques to ensure equipment reliability and maintainability.

maintenance information system – A means by which field data are converted into information so that maintenance can determine work needed, control the work, and measure the effectiveness of the work done.

maintenance work order (MWO) – A formal document for controlling planned and scheduled work.

maintenance work request (MWR) – An informal document for requesting unscheduled or emergency work. Also called a job ticket or job request.

major repairs – Extensive, nonroutine, scheduled repairs requiring the deliberate shutdown of equipment, the use of a repair crew (possibly covering several elapsed shifts), significant materials, rigging, and, if needed, the use of lifting equipment.

material coordinator – The person assigned to planning staff who is responsible for the procurement of all materials required for planned work.

mean time before failure (MTBF) – The average time between replacements of a specific component on a designated type of equipment. Also referred to as the life-span of a component. Extended MTBF indicates successful actions in extending component life-span.

minor repairs – Repairs usually performed by one person using hand tools, few parts, and usually completed in less than 2 hours.

multicraft – Requires personnel to possess more than one craft.

multiskilled – Maintenance that blends necessary craft skills allowing personnel to do a job from start to finish. Also refers to flexible trades, cross-trades, and multicraft maintenance.

nondestructive testing – The use of testing technologies to detect cracks, flaws, or porosity in components, structures, frames, or components. Techniques include magnetic particle testing, liquid dye penetrant, ultrasonic testing, eddy current testing, and radiographic testing. See also *predictive maintenance.*

objective – The principal purpose for the existence of each line department (e.g., maintenance) or staff department (e.g., purchasing) and the roles they must play to ensure that the plant production strategy is achieved.

oil analysis – Spectrometric analysis to identify wear particles from the equipment and physical tests to provide information about the lubricant being evaluated.

on-condition status – After the discovery of a potential failure, equipment is left in operation on condition that it can continue to perform its intended function. Equipment condition is monitored carefully during this period to preclude a sudden deterioration to a functional failure.

overhaul – An equipment condition in which a unit of equipment must be removed from service and subjected to inspection, tear-down, and repair of the total unit to restore it to effective operating condition in accordance with current design specifications. See also *rebuild.*

performance indices – Ratios which convey short-term accomplishments and long-term trends against desired standards.

periodic maintenance – Maintenance actions carried out at regular intervals. Intervals may be fixed (e.g., every 6 months) or variable (e.g., every 4,500 operating hours).

P–F curve – A down parabolic shaped curve denoting the deterioration of equipment condition from the discovery of a potential failure (P) to a functional failure (F).

P–F interval – The elapsed time between the discovery of a potential failure (P) until a functional failure (F) occurs if no corrective action is taken.

physical testing – Physical testing as with an oil sample to reveal how the lubricant performs. If lubricant contamination is found, the lubricant is changed before further wear occurs.

planning – Determination of resources needed and the development of anticipated actions necessary to perform a scheduled major job.

policies – Management guidelines for the development of field procedures to ensure achievement of plant profitability. See also *guidelines.*

predictive maintenance – Techniques to predict wear rate, determine state of deterioration, monitor condition, or predict failure.

preventive maintenance (PM) services – Performance of services to avoid premature equipment failure and extend equipment life. Specifically, equipment inspection, testing, and condition monitoring to ensure the early detection of equipment deficiencies and lubrication, cleaning, adjusting, calibration, and minor component replacements to extend equipment life.

principles – Logic, common sense, proven procedures, or essential rules on which plant operation must be based.

proactive maintenance – The application of investigative and corrective technologies to reduce failures, improve equipment performance, and extend equipment life. The following analytical tools are associated with proactive maintenance: root cause failure analysis, failure modes and effects analysis, and risk-based inspections. Also, the intensive application of positive, aggressive maintenance steps to actively defeat potential failures.

procedure – The day-to-day method for carrying out elements of the maintenance program such as job assignment and work control; established steps to carry out a task.

production strategy – The manager's plan for achieving plant profitability. Also called an operating plan or business plan.

productivity – The percentage of time that maintenance personnel are at the work site, with their tools, performing productive work during a scheduled working period.

program – A plan under which action can be taken toward a goal.

program (maintenance) – The interaction of the total plant population as they request or identify work; classify it to determine the best reaction; then plan (as required), schedule, assign, control, and measure the resulting work; and finally, assess overall accomplishment against goals such as performance standards and budgets.

priority – The relative importance of a single job in relationship to other jobs, operational needs, safety, equipment condition, and so on, and the time within which the job should be completed.

project – A definitive objective to build or construct a new capital entity. The installation or modification of major equipment.

project work – Actions such as construction, equipment modification, installation, or relocation to gain economic advantage; replacement of worn, damaged, or obsolete equipment; satisfaction of a safety requirement; attainment of additional operating capacity or the meeting of a basic need. Work is usually capital-funded.

purchase order (PO) – The authorized document for obtaining direct charge materials or services from vendors or contractors.

quality standard – A standardized procedure for accomplishing a major maintenance task in the best way.

quantity standard – The standard resources required to meet the prescribed quality standard.

random sampling – A statistical technique of data gathering based on the laws of probability. Observations made at random times yield a picture of what happens most of the time. More observations provide a more reliable picture.

rebuild – The repair of a component to restore it to serviceable condition in accordance with current design specifications. See also *overhaul.*

relocate – To move fixed equipment to a different location.

reliability – The overall condition of production equipment measured by the extended life-span of internal components.

reliability engineering – Actions taken through the use of information, field experience, and engineering techniques to design or redesign equipment in order to reduce or eliminate faults that imperil equipment reliability.

reliability centered maintenance (RCM) – A strategy for achieving maximum equipment reliability and extended life at the least cost. Implementation identifies specific equipment functions in their exact operating context. Then equipment performance standards are identified for each function and failures defined when performance standards are not met. Based on the consequences of failures, a maintenance program featuring condition-monitoring techniques is applied to identify potential failures (equipment is starting to fail) accurately and quickly to preclude their deterioration to functional failure (equipment no longer operates) levels. Thus, equipment life is extended and the consequences of functional failures are reduced or avoided. See also *P–F curve, P–F interval, on-condition status.*

repair history – A chronological record of significant repairs made on key equipment used to spot chronic, repetitive problems, failure patterns, and component life-span, which in turn identifies corrective actions and helps forecast component replacements.

repetitive maintenance – Maintenance jobs that have a known labor and material content and occur at a regular interval.

reposition – To move mobile equipment to a new working location.

risk-based inspection – Guides decisions in selecting equipment that possesses the greatest risk of failure if inspections are not done by ranking the equipment according to its probability and consequences of failure. Then the most essential inspection services are contrasted against the raking of equipment to allow the best allocation of resources to meet the basic inspection needs. Risk-based inspection applies primarily to piping.

risk priority number (RPN) – In a failure modes and effects analysis, an RPN is assigned to the failures related to equipment being studied to establish the priority of corrective actions. RPN = severity × occurrence × detection capability. See also *failure modes and effects analysis.*

root cause failure analysis (RCFA) – Actions to find why a failure happened and determine corrective actions to eliminate the failure or reduce its impact.

routine maintenance – Maintenance or services performed consistently in the same manner.

schedule compliance – The effectiveness with which an approved schedule was carried out. Reported as the percentage of jobs completed versus those scheduled.

scheduling – Determination of the best time to perform a planned maintenance job to appreciate operational needs for equipment or facilities and the best use of maintenance resources.

self-directed team – A team of personnel in which each member shares equally in decision making, control, conduct of work, and the accountability for results.

specifications – Technical definition of equipment configuration or performance requirements to meet intended utilization of equipment or materials.

standard – A goal or ideal target to be met. Quality standards prescribe the end product. Quantity standards prescribe the amount of resources required to carry out specific work under normal conditions.

standard operating procedures – A written procedure used to ensure reasonable uniformity each time a significant task is performed.

standing work order – A reference number used to identify a routine, repetitive action.

stock issue card – The authorized accounting document for making stock material withdrawals or returns.

strategy – A global, corporate, or plantwide plan to secure a major objective, such as the successful implementation of total productive maintenance.

system – Standard procedures to accomplish tasks in an organized way.

team (maintenance) – A group of maintenance personnel with complementary skills who are mutually accountable for a common goal of effective maintenance and committed to quality performance.

team coordinator – The focal point of work control within a team. Often rotated to ensure that team orientation is preserved.

time based – Repetitive actions taken based on the passage of a specific time period or the accumulation of a certain number of operating hours or unit of product (e.g., tons).

time card – The authorized accounting document for reporting the use of labor data.

total productive maintenance (TPM) – Productive maintenance carried out by all employees through small-group activities. TPM is preventive maintenance performed on a plantwide basis.

type – All equipment of the same kind: conveyors, pumps, haulage trucks, loaders, and so forth.

unit – One piece of equipment of a specific type (e.g., conveyor 006).

utilization – The percentage of time that a maintenance crew is available to perform productive work during a scheduled working period or shift.

verbal orders – A means of assigning emergency work when reaction time does not permit preparation of a work order document.

work force – The personnel who carry out the maintenance work load.

work load – The essential work to be performed by maintenance and the conversion of this data into a work force of the proper size and craft composition, working at reasonable productivity, to ensure the maintenance program is carried out effectively.

work order system – A communications system by which maintenance work is requested or identified, classified, planned, scheduled, assigned, and controlled.

Bibliography

Allen, T. 2006. RCM, the Navy Way for Optimal Submarine Operations. RCM Forum in Las Vegas, NV.

Barlas, S. 1992. U.S. companies feverishly seek ISO 9000 registration. *Managing Automation*, March.

Benson, T.E. 1991. Quality goes international. *Industry Week*, August 19.

Berkman, B.N. 1989. European companies join the quality crusade. *Electronic Business*, October 16.

Bittel, L.R. 1980. *What Every Supervisor Should Know*. New York: McGraw-Hill.

Bogan, C.E., and English, M.J. 1994. *Benchmarking for Best Practices*. New York: McGraw-Hill.

Boyd, B.B. 1976. *Management-Minded Supervision*. New York: McGraw-Hill.

Buzzell, R.D., and Gale, B.T. 1987. *The PIMS Principles: Linking Strategy to Performance*. New York: Free Press.

Costanzo, A. 1992. U.S. corporate executive knowledge of ISO 9000 lacking. *Quality*, September.

Crosby, P. 1979. *Quality Is Free*. New York: McGraw-Hill.

Daniels College of Business. 1995. *The Quality Panorama*. University of Denver, Colorado.

Deming, W.E. 1982. *Quality, Productivity, and Competitive Position*. Cambridge, MA: Center for Advanced Engineering Study.

Feigenbaum, A.V. 1990. The criticality of quality and the need to measure it. *Financier*, October.

Fogel, G. 1995. Upgrading oil analysis to tribology. *Maintenance Technology*, October.

Fuchs, S.J. 1993. *Complete Building Equipment Maintenance Desk Book*. Englewood Cliffs, NJ: Prentice Hall.

Gilley, J.W., and Eggland, S.A. 1989. *Principles of Human Resource Development*. Reading, MA: Addison-Wesley.

Graham, J.F. 1991. ISO-9000 certification: Maintenance's role. *Plant Services*, November.

Hanks, K. 1991. *Motivating People*. Los Altos, CA: Crisp Publications.

Hartman, E.H. 1992. *Successfully Installing TPM in a Non-Japanese Plant*. Allison Park, PA: TPM Press.

Haynes, M.E. 1989. *Project Management—From Idea to Implementation*. Menlo Park, CA: Crisp Publications.

Hutchens, S. 1991. Facing the ISO-9000 challenge. *Compliance Engineering*, Fall.

International Organization for Standardization. 1992. *ISO 9000: International Standard for Quality Management*, 2nd ed. Geneva, Switzerland: ISO.

Johnson, G. 1991. American firms face challenge of meeting new quality guidelines. *Denver Business*, June/July.

Johnson, P.L. 1989. *Keeping Score: Strategies and Tactics for Winning the Quality War*. New York: Harper Business.

———. 1993. *ISO 9000—Meeting the New International Standards*. New York: McGraw-Hill.

Johnson, P.L., and Kantner, R. 1992. *All About ISO 9000*. Southfield, MI: Perry L. Johnson.

———. 1992. *How to Write Your ISO 9000 Quality Manual*. Southfield, MI: Perry L. Johnson.

Jones, E.K. 1996. Contract maintenance strategies. *Maintenance Technology*, February and March.

Judson, A.S. 1966. *A Manager's Guide to Making Changes*. New York: John Wiley & Sons.

Juran, J.M. 1951. *Quality Control Handbook*. New York: McGraw-Hill.

Juran, J.M., and Gryna, F.M. 1988. *Juran's Quality Control Handbook*. New York: McGraw-Hill.

Kelly, A. 1989. *Maintenance and Its Management*. South Hampton, England: Ashford Press.

Kerzner, H. 1995. *Project Management*, 5th ed. New York: Van Nostrand-Reinhold.

Kiplinger Washington Letter. 1992. 69(43), October 23.

Lamprecht, J.L. 1992. *ISO 9000: Preparing for Registration*. New York: Marcel Dekker.

Maddux, R.B. 1992. *Team Building*. Los Altos, CA: Crisp Publications.

Maggard, B.N. 1992. *TPM That Works: Theory and Design of Total Productive Maintenance*. Allison, PA: TPM Press.

Maynard, H.B. 1971. *Industrial Engineering Handbook*, 3rd ed. New York: McGraw-Hill.

Mercer, D. 1991. Total quality management: Key quality issues. In *Global Perspectives on Total Quality*. New York: Conference Board.

Morrow, L.C. 1957. *Maintenance Engineering Handbook*. New York: McGraw-Hill.

Moubray, J. 1996. Redefining maintenance. *Maintenance Technology*. March, April, and May.

Moubray, J.M. 1994. *Reliability-Centered Maintenance*. Oxford: Butterworth-Heinemann.

Nakajima, S. 1988. *Total Productive Maintenance*. Cambridge, MA: Productivity Press.

———. 1989. *TPM Development: Implementing TPM*. Portland, OR: Productivity Press.

Niese, A. 1992. ISO 9000: International Standard for Quality. *Electronic News*, November.

Nowlan, F.S., and Heap, H.F. 1978. *Reliability-Centered-Maintenance*. Washington, DC: Office of Assistant Secretary of Defense.

Okes, D.W. n.d. *Overview of the ISO 9000 Quality Systems Standards*. Aplomet.

Placek, C. 1991. Agreement on standards, testing, and certification outpacing agreement on many other issues facing European community. *Quality*, October.

———. 1992. The ISO 9000 edge. *Quality*, January.

Resnikoff, H.L. 1978. *Mathematical Aspects of Reliability-Centered-Maintenance*. Los Altos, CA: Dolby Access Press.

Robinson, C.J., and Ginder, A.P. 1995. *Implementing TPM, The North American Experience*. Portland, OR: Productivity Press.

Ross, J.E. 1995. *Total Quality Management*. Delray Beach, FL: St. Lucie Press.

Scholtes, P.R. 1992. *Team Handbook*. Madison, WI: Joiner Associates.

Stratton, J.H. 1992. What is the Registrar Accreditation Board? *Quality Progress*, January.

Tiratto, J. 1991. Preparing for EC 1992 in the US through quality system registration. *Computer*, April.

Tomlingson, P.D. 1986. *Maintenance Management for First-Line Supervisors*, Vols. 1–5. Barrington, IL: TPC Training Systems.

———. 1992. *Maintenance Management Information Systems*. Denver, CO: Kistler Press.

———. 1993. *Effective Maintenance—Key to Profitability*. New York: Van Nostrand-Reinhold.

———. 1994. *Mine Maintenance Management*, 9th ed. Dubuque, IA: Kendall/Hunt Publishing.

———. 1995. *Industrial Maintenance Management*, 9th ed. Dubuque, IA: Kendall/ Hunt Publishing.

Webb, N.M. 1991. Flies in the soup. *Quality in Manufacturing*, July/August.

Young, K. 1995. Integrating predictive maintenance. *Maintenance Technology Magazine*, June.

Index

NOTE: *f.* indicates figure; *t.* indicates table.

D

E

N

About the Author

For more than 35 years, **Paul D. Tomlingson** has provided management consulting services in the design, implementation, and evaluation of industrial maintenance management programs for heavy industry. Tomlingson is a 1953 graduate of West Point. In addition to a BS degree in engineering, he holds an MA degree in government and an MBA degree, both from the University of New Hampshire.

Tomlingson has published more than 100 trade journal articles on maintenance management, principally in *Engineering and Mining Journal, Coal Age, Mining Engineering, Plant Engineering, Production Engineering, and Maintenance Technology*. He is the author of seven books: *Equipment Management: Key to Equipment Reliability and Productivity in Mining (2010), Mine Maintenance Management Reader* (2007), *Equipment Management: Breakthrough Maintenance Management Strategy for the 21st Century* (1998), *Mine Maintenance Management* (1994), *Industrial Maintenance Management* (1996), *Maintenance Management Information Systems* (1993), and *Effective Maintenance: The Key to Profitability* (1993). He is also the author of Technical Publishing Company's five-volume training text, *Maintenance Management for First-line Supervisors* (1981), and a contributing author to Prentice-Hall's *Complete Equipment Maintenance Desk Book*, Second Edition (1992).

In addition to the presentation of his own public and on-site seminars, Tomlingson has presented seminars at the University of Wisconsin; Hofstra University (New York); the University of Denver (Colorado); Concordia University (Quebec); the Colorado School of Mines; Virginia Tech; the International Institute for Learning, Inc. (New York); the International Institute for Research (Dubai, UAE); McGill University (Quebec); national and regional plant engineering shows; and SME conferences.

Tomlingson has been listed in *Who's Who in the West*, is a member of SME, and is a founding member of the Mining Industry Council for Equipment Management.